Research and Perspectives in Alzheimer's Disease

For further volumes:
http://www.springer.com/series/1175

Mathias Jucker • Yves Christen
Editors

Proteopathic Seeds and Neurodegenerative Diseases

Springer

Editors
Mathias Jucker
University of Tübingen
Department of Cellular Neurology
Hertie Institute f. Clin. Brain Research
Tübingen, Germany

Yves Christen
Fondation IPSEN
Boulogne-Billancourt Cedex
France

ISSN 0945-6066
ISBN 978-3-642-35490-8   ISBN 978-3-642-35491-5 (eBook)
DOI 10.1007/978-3-642-35491-5
Springer Heidelberg New York Dordrecht London

Library of Congress Control Number: 2013935112

Printed on acid-free paper

Springer is part of Springer Science+Business Media (www.springer.com)

Foreword

The misfolding and aggregation of specific proteins is an early and obligatory event in many of the age-related neurodegenerative diseases of humans. The initial cause of this pathogenic cascade and the means whereby disease spreads through the nervous system, remain uncertain.

A recent surge of research, first instigated by pathologic similarities between prion diseases and Alzheimer's disease, increasingly implicates the conversion of disease-specific proteins into an aggregate-prone β-sheet-rich state as the prime mover of the neurodegenerative process. This prion-like corruptive protein templating or seeding now characterizes such clinically and etiologically diverse neurological disorders as Alzheimer's disease, Parkinson's disease, Huntington's disease, amyotrophic lateral sclerosis, and frontotemporal lobar degeneration. Understanding the misfolding, aggregation, trafficking and pathogenicity of the affected proteins could therefore reveal universal pathomechanistic principles for some of the most devastating and intractable human brain disorders.

The aim of the meeting entitled like this volume "proteopathic seeds and neurodegenerative diseases" held in February 2012 in Paris is to gather state-of-the-art knowledge in this rapidly moving research field. This volume is also intended to further stimulate the translation of this emerging common pathomechanism into the treatment and prevention of neurodegenerative diseases. Because disease-specific protein aggregates can be detected many years to decades before clinical symptoms manifest, proteopathic seeds may become early and disease-specific biomarkers. The volume also reveals that the prion-like transformation of protein structure can in some circumstances be beneficial and is exploited by the healthy organism to store and transmit information at the molecular level.

It is time to accept that the prion concept is no longer confined to prionoses but is a promising concept for the understanding and treatment of a remarkable variety of diseases that afflict primarily our aging society.

Tübingen, Germany   Mathias Jucker
Boulogne-Billancourt Cedex, France   Yves Christen

Contents

List of Contributors

Elodie Angot Neuronal Survival Unit, Wallenberg Neuroscience Center, Lund University, Lund, Sweden

Anne Bertolotti MRC Laboratory of Molecular Biology, Cambridge, UK

Patrik Brundin Neuronal Survival Unit, Wallenberg Neuroscience Center, Lund University, Lund, Sweden

Center for Neurodegenerative Science, Van Andel Research Institute, Grand Rapids, MI, USA

Florence Clavaguera Institute of Pathology, University of Basel, Basel, Switzerland

Irina Derkatch Center for Neurobiology and Behavior, Howard Hughes Medical Institute, Columbia University, New York, NY, USA

Christopher Dunning Neuronal Survival Unit, Wallenberg Neuroscience Center, Lund University, Lund, Sweden

C. Duran-Aniotz Mitchell Center for Alzheimer's Disease and Related Brain Disorders, Department of Neurology, University of Texas Houston Medical School, Houston, TX, USA

Bahareh Eftekharzadeh Department of Cellular Neurology, Hertie Institute for Clinical Brain Research, University of Tübingen, Tübingen, Germany

DZNE, German Center for Neurodegenerative Diseases, Tübingen, Germany

Yvonne S. Eisele Department of Cellular Neurology, Hertie Institute for Clinical Brain Research, University of Tübingen, Tübingen, Germany

DZNE, German Center for Neurodegenerative Diseases, Tübingen, Germany

Sarah K. Fritschi Department of Cellular Neurology, Hertie Institute for Clinical Brain Research, University of Tübingen, Tübingen, Germany

DZNE, German Center for Neurodegenerative Diseases, Tübingen, Germany

Michel Goedert MRC Laboratory of Molecular Biology, Cambridge, UK

Jason Greenwald ETH Zurich, Physical Chemistry, ETH Honggerberg, Zurich, Switzerland

Tsuyoshi Hamaguchi Department of Cellular Neurology, Hertie Institute for Clinical Brain Research, University of Tübingen, Tübingen, Germany

DZNE, German Center for Neurodegenerative Diseases, Tübingen, Germany

Götz Heilbronner Department of Cellular Neurology, Hertie Institute for Clinical Brain Research, University of Tübingen, Tübingen, Germany

DZNE, German Center for Neurodegenerative Diseases, Tübingen, Germany

Mathias Jucker Department of Cellular Neurology, Hertie Institute for Clinical Brain Research, University of Tübingen, Tübingen, Germany

DZNE, German Center for Neurodegenerative Diseases, Tübingen, Germany

Eric R. Kandel Center for Neurobiology and Behavior, Howard Hughes Medical Institute, Columbia University, New York, NY, USA

Franziska Langer Department of Cellular Neurology, Hertie Institute for Clinical Brain Research, University of Tübingen, Tübingen, Germany

DZNE, German Center for Neurodegenerative Diseases, Tübingen, Germany

Peter T. Lansbury Jr Center for Neurologic Diseases, Brigham and Women's Hospital, Boston, MA, USA

Giusi Manfredi Department of Cellular Neurology, Hertie Institute for Clinical Brain Research, University of Tübingen, Tübingen, Germany

DZNE, German Center for Neurodegenerative Diseases, Tübingen, Germany

R. Morales Mitchell Center for Alzheimer's Disease and Related Brain Disorders, Department of Neurology, University of Texas Houston Medical School, Houston, TX, USA

I. Moreno-Gonzalez Mitchell Center for Alzheimer's Disease and Related Brain Disorders, Department of Neurology, University of Texas Houston Medical School, Houston, TX, USA

Amudha Nagarathinam Department of Cellular Neurology, Hertie Institute for Clinical Brain Research, University of Tübingen, Tübingen, Germany

DZNE, German Center for Neurodegenerative Diseases, Tübingen, Germany

Elias Pavlopoulos Center for Neurobiology and Behavior, Howard Hughes Medical Institute, Columbia University, New York, NY, USA

Stanley B. Prusiner Institute for Neurodegenerative Diseases, University of California, San Francisco, CA, USA

Department of Neurology, University of California, San Francisco, CA, USA

Nolwen L. Rey Neuronal Survival Unit, Wallenberg Neuroscience Center, Lund University, Lund, Sweden

Roland Riek ETH Zurich, Physical Chemistry, ETH Honggerberg, Zurich, Switzerland

Claudio Soto Mitchell Center for Alzheimer's Disease and Related Brain Disorders, Department of Neurology, University of Texas Houston Medical School, Houston, TX, USA

Jennifer A. Steiner Center for Neurodegenerative Science, Van Andel Research Institute, Grand Rapids, MI, USA

Markus Tolnay Institute of Pathology, University of Basel, Basel, Switzerland

Robert Tycko Laboratory of Chemical Physics, National Institute of Diabetes and Digestive and Kidney Diseases, National Institutes of Health, Bethesda, MD, USA

Lary Walker Yerkes National Primate Research Center, Emory University, Atlanta, GA, USA

Department of Neurology, Emory University, Atlanta, GA, USA

Gunilla T. Westermark Department of Medical Cell Biology, Uppsala University, Uppsala, Sweden

Per Westermark Department of Immunology, Genetics and Pathology, Uppsala University, Uppsala, Sweden

Widening Spectrum of Prions Causing Neurodegenerative Diseases

Stanley B. Prusiner

Abstract The field of prion biology still seems to be in its infancy. Over the past three decades, there has been a steady accumulation of evidence that each neurodegenerative disease is caused by a particular protein that becomes a prion. As with the prion diseases caused by the aberrant prion protein (PrP^{Sc}), amyloid deposits in other neurodegenerative disorders were found to have the same protein as that identified by molecular genetic studies of patients with inherited neurodegeneration. While the number of prions identified in mammals (now at more than half a dozen) and in fungi (now more than ten) will undoubtedly continue to expand, we have no idea about prions in all the other phylogeny. The mammalian prions composed of PrP, Aβ, tau, α-synuclein, SOD1 and huntingtin all cause distinct neurodegenerative diseases. In each of these disorders, the respective mammalian proteins adopt a β-sheet–rich conformation that readily oligomerizes and becomes self-propagating. The oligomeric states of mammalian prions are thought to be the toxic forms, and assembly into larger polymers such as amyloid fibrils seems to be a mechanism for minimizing toxicity. To date, there is not a single medication that halts or even slows a neurodegenerative disease caused by prions. This may be a bellwether of the unique pathogenic mechanisms that feature in each of the prion diseases and of the urgent need to develop informative molecular diagnostics and effective antiprion therapeutics.

Following the experimental transmission of the human disorders kuru and Creutzfeldt-Jakob disease (CJD) to apes and monkeys, the search intensified for a slow-acting virus causing the analogous disease in sheep and goats, called scrapie (Gajdusek 1977). As preparations from brains of scrapie-infected hamsters were enriched for

S.B. Prusiner (✉)
Institute for Neurodegenerative Diseases, University of California, 675 Nelson Rising Ln, Room 318, Box 0518, San Francisco, CA 94143, USA

Department of Neurology, University of California, San Francisco, CA 94143, USA
e-mail: stanley@ind.ucsf.edu

M. Jucker and Y. Christen (eds.), *Proteopathic Seeds and Neurodegenerative Diseases*, Research and Perspectives in Alzheimer's Disease, DOI 10.1007/978-3-642-35491-5_1, © Springer-Verlag Berlin Heidelberg 2013

infectivity, evidence for an essential protein emerged but no similar data for a nucleic acid could be generated. To the consternation of many, I introduced the term "prion" in order to distinguish the proteinaceous infectious particles causing scrapie and CJD from viruses (Prusiner 1982). Despite numerous attempts to demonstrate a scrapie-specific nucleic acid, none could be found and, therefore, those investigations were eventually abandoned (Safar et al. 2005).

Soon thereafter, a protein with a molecular weight of 27–30 kDa was found in purified fractions containing high levels of scrapie infectivity (Bolton et al. 1982; McKinley et al. 1983). This protein was designated prion protein 27–30 (or PrP 27–30). Scrapie infectivity and PrP 27–30 were found at the top of sucrose gradients, indicating that some of the infectivity was likely to be very small in accord with the small ionizing radiation target size, whereas the majority of the infectivity sedimented to the bottom (Alper et al. 1966; Bellinger-Kawahara et al. 1988; Prusiner et al. 1983). The rod-shaped structures at the bottom of the gradients were shown to be composed of PrP 27–30 and to stain with Congo red dye, establishing the amyloid nature of these large aggregates (Prusiner et al. 1983). That discovery suggested that other amyloid deposits in disorders such as Alzheimer's and Parkinson's diseases might be composed of causative proteins, i.e., prions (Prusiner 1984).

Plaques, Tangles, and Inclusions

Beginning in the mid-1980s, the proteins in plaques, tangles, and intracellular bodies of the brains of patients who died of neurodegeneration were identified through purification or immunostaining (Brion et al. 1985; Glenner and Wong 1984a; Grundke-Iqbal et al. 1986; Kosik et al. 1986; Spillantini et al. 1997; Wood et al. 1986). Each of these proteins was found to aggregate into fibrils under some conditions and to form amyloids. Those same proteins were also identified by genetic studies in which the mutant genes causing familial forms of neurodegeneration were found (Goate et al. 1991; Hsiao et al. 1989; Hutton et al. 1998; Polymeropoulos et al. 1997). Subsequently, expression of these mutant genes in transgenic (Tg) mice was shown to recapitulate many aspects of the human illnesses (for review, see Prusiner 2001). Despite these similarities with scrapie and CJD, many investigators still prefer to think about the other neurodegenerative diseases as unrelated, since Gajdusek, Gibbs and their collaborators were unable to transmit Alzheimer's disease (AD) to apes and monkeys (Godec et al. 1991; Goudsmit et al. 1980).

Looking back, attempts to demonstrate the transmissibility of Alzheimer's and Parkinson's diseases to apes and monkeys were made long before the Aβ peptide and α-synuclein were isolated, so neither could be used as biomarkers. Much later, Ridley and Baker demonstrated transmission of disease to marmosets, as measured by Aβ cerebral amyloidosis, after intracerebral injections of AD brain homogenates (Baker et al. 1994; Ridley et al. 2006). The incubation times in those studies exceeded 3.5 years, making confirmation of such experiments impractical. Later, Tg mouse studies would supplant the use of nonhuman primates.

Table 1 Evidence for prions causing many different neurodegenerative diseases

Prion diseases	Precursor proteins	Prion forms	Protein deposits	Self-propagation in mammals	Self-propagation in cells
CJD/scrapie	PrP^C	PrP^{Sc}	PrP plaques	Inoc wt & Tg mice	N2a, GT1
Alzheimer's	APP	Aβ	Aβ plaques	Inoc Tg(ΔAPP) mice	
Tauopathies (FTD, PSP, Pick's, CTE)	tau	tau aggregates	NFTs, Pick bodies	Inoc Tg(HuTau), inoc Tg(HuTau,P301S) and inducible Tg(HuTau, ΔK280) mice	C17.2, HEK293
Parkinson's	α-Synuclein	α-Synuclein aggregates	Lewy bodies	Lewy bodies in grafts and inoc Tg (HuSNCA,H53T) mice	Primary mouse hippocampal neurons
fALS	ΔSOD1, ΔTDP43, ΔFus	ΔSOD1 aggregates	Bunina bodies		N2a, HEK

inoc, inoculum; Δ, mutant.

Aβ Prions

In a series of incisive experiments, Mathias Jucker and Lary Walker collaborated on the transmission of Aβ prions to weanling Tg(APP23) mice expressing a mutant human amyloid precursor protein (APP). Much of their work was performed with brain homogenates from aged Tg(APP23) mice that spontaneously developed Aβ amyloid plaques. The inoculum prepared from these old mice dramatically accelerated the deposition of nascent Aβ amyloid, and immunoabsorption of Aβ in the inoculum prevented the accelerated deposition of Aβ (Table 1; Meyer-Luehmann et al. 2006). These investigators also reported transmission of Aβ prions after intraperitoneal injections (Eisele et al. 2009). More recently, Claudio Soto and colleagues as well as Jucker and Walker have reported that brain homogenates from patients who died of AD could also transmit disease to Tg mice and rats expressing wild-type (wt) human APP, respectively (Morales et al. 2012; Rosen et al. 2012).

Using bigenic mice expressing mutant APP and luciferase under control of the murine *Thy-1.2* and *Gfap* promoters, respectively, my colleagues and I found brain homogenates from aged Tg(APP23) mice transmitted Aβ prions, as reflected by increased bioluminescence and Aβ deposits (Watts et al. 2011). Subsequently, we found that fractions highly enriched for Aβ produced an increase in bioluminescence after approximately 5 months, which was ~6 weeks more rapidly than that seen with the crude brain homogenates. Most important, synthetic Aβ peptides became prions during polymerization into amyloid, establishing that some hypothetical contaminate was not responsible for the change in bioluminescence (Stöhr et al. 2012).

Remarkably, Heiko Braak not only described the spreading of Aβ amyloid but also showed the concurrent deposition of neurofibrillary tangles (NFTs) composed of the tau protein (Braak et al. 1996; Braak and Del Tredici 2011). Recent studies have traced the spread of tau prions using functional magnetic resonance imaging (fMRI) intrinsic connectivity analysis (Seeley et al. 2009). Such a spreading of prions was described much earlier for scrapie prions in ovines and rodents (Hadlow et al. 1974; Kimberlin and Walker 1979; Scott et al. 1992; Taraboulos et al. 1992; Tatzelt et al. 1999).

Inherited CJD and AD

As with familial (f) CJD, fatal familial insomnia (FFI), and Gerstmann-Sträussler-Scheinker (GSS) disease, inherited AD is caused by mutations in the gene that encodes the protein found in amyloid plaques. The recognition that 10–15% of CJD cases are familial led to the suspicion that genetics plays a role in this disease. Subsequently, transmission of fCJD to apes and monkeys was reported (Masters et al. 1981; Roos et al. 1973). More than 40 different mutations of the PrP gene have been identified, of which 35 are point mutations and the remainder are octarepeat expansions or deletions (see reviews: Mead 2006; van der Kamp and Daggett 2009); five mutations have been genetically linked to inherited human prion diseases (Hsiao et al. 1989). Virtually all cases of GSS and FFI appear to be caused by germline mutations in the PrP gene. The brains of patients with inherited prion disease contain infectious prions that have been transmitted to experimental animals.

In fAD, missense mutations have been found in APP and the presenilins (Hardy and Selkoe 2002; St. George-Hyslop 1999). The amyloid fibrils in AD were found to contain the Aβ peptide (Glenner and Wong 1984b; Masters et al. 1985) that is cut from the larger APP. Many cases of fAD are caused by mutations in APP or in enzymes that produce Aβ (Goate et al. 1991; Schellenberg et al. 1992; St. George-Hyslop et al. 1992). Subsequently, Tg mice expressing mutant human APP exhibited amyloid plaques composed of the Aβ peptide (Games et al. 1995; Hsiao et al. 1996).

Tau Prions

The role of the tau protein in the pathogenesis of AD was resolved when mutations in the tau gene were found to cause heritable tauopathies, including familial frontotemporal dementia (FTD), inherited progressive supranuclear palsy (PSP), and Pick's disease but not fAD (Hong et al. 1998; Hutton et al. 1998; Rademakers and Hutton 2007). Aggregates formed from truncated recombinant human tau composed of residues 242–364 were shown to enter C17 cells and seed the polymerization of endogenous tau (Frost et al. 2009). These important studies were extended using HEK293 cells expressing full-length wt human tau(2N4R) as well as truncated and mutant human tau(P301S) (Guo and Lee 2011). Michel Goedert and his colleagues described the transmission of mutant tau(P301S) prions produced in Tg mice to recipient mice expressing wt human tau (Clavaguera et al. 2009). After approximately 6 months, the inoculated Tg mice showed wt tau aggregates that had spread from the site of inoculation to neighboring regions.

Like the Aβ amyloid plaques of AD, the NFTs also spread along neuroanatomical pathways (Braak and Braak 1995). Presumably, the tau prions spread transsynaptically as they move from one neuron to another. Recent Tg mouse models expressing human tau in the entorhinal cortex show spread along neuroanatomically defined pathways to hippocampal pyramidal neurons, especially in CA1 and dentate gyrus granule cells (de Calignon et al. 2012; Liu et al. 2012; Polymenidou and Cleveland 2012).

A New Psychiatry

The tauopathies sit at an interface between psychiatry and neurology. A minority of FTD patients harbor tau gene mutations whereas the vast majority do not (Rademakers and Hutton 2007). Importantly, ablation of the tau gene in Alzheimer's mouse models expressing mutant APP ameliorates disease (Roberson et al. 2007). Often, psychiatrists see FTD patients for several years before they refer them to neurologists with the diagnosis of AD. Tau lesions in the frontal lobes can produce behavioral disinhibition, apathy, inappropriate social interactions, depression, insomnia, and executive dysfunction (Rabinovici and Miller 2010). Later, semantic dementia, drug abuse, alcoholism, and sometimes suicide are seen. Different phenotypes, including PSP, corticobasal degeneration, and Pick's disease, may be due to different strains of tau prions producing lesions in different neuronal circuits (Seeley et al. 2009).

In an unanticipated finding, contact sport athletes, including boxers and football and hockey players, have symptoms similar to those of FTD patients. Numerous NFTs have been found in the frontal lobes of contact sports athletes (Corsellis et al. 1973), some of whom committed suicide. While the first football player in whom a tauopathy was recognized played professional football for 16 seasons and died at age 50 (Omalu et al. 2005), other players in high school and/or college have also been identified with tauopathies (McKee et al. 2009). Recently, a 27-year-old Marine, who suffered multiple episodes of traumatic brain injury (TBI) during explosions of roadside bombs in Iraq, was found to suffer from a tauopathy. Following the diagnosis of posttraumatic stress disorder (PTSD), he was honorably discharged, divorced his wife, and became an alcoholic. Eight months after honorable discharge, he committed suicide by hanging; at autopsy, numerous NFTs were found in his frontal lobes (Omalu et al. 2011).

Late-Onset Neurodegeneration

An important characteristic shared by many of the neurodegenerative diseases is the finding that 10–20% are inherited and more than 80% are sporadic. In addition to CJD, this relationship has been found for Alzheimer's, Parkinson's, the tauopathies, and amyotrophic lateral sclerosis (ALS) but not for Huntington's, which is always inherited. Although most mutant proteins are expressed early in embryogenesis, the vast majority of neurodegenerative diseases are late-onset illnesses, including fCJD. This finding argues that some event occurs with aging that renders a disease-specific protein pathogenic. I argue that this event involves the refolding of the etiologic protein into a prion state, which is able to support the self-propagation of more prions (Prusiner 1989). Importantly, many of the mutant proteins causing heritable neurodegenerative diseases are the same as those found in disease-specific amyloid deposits, such as plaques, NFTs, and Lewy bodies (Table 1; for review, see Prusiner 2001).

One fascinating insight into the control of the timing of disease onset comes from studies of ~35 fCJD families with expanded PrP octarepeat regions (Croes et al. 2004; Stevens et al. 2009). Wild-type PrP^C contains four octarepeats; one to four additional octarepeats cause late-onset fCJD with a mean age of 62 years whereas five to eight additional octarepeats cause early-onset fCJD with a mean age of 32 years. In other words, the fifth additional octarepeat decreases the age of disease onset by three decades. This profound transition in pathogenesis was found to correlate with a shift in Cu^{2+} binding to PrP: a high Cu^{2+} occupancy state changes to a low one (Stevens et al. 2009). Titrations showed that one Cu^{2+} ion bound to each histidine in the high occupancy state; each of the four octarepeats contains a single histidine. In the low occupancy state, a single Cu^{2+} ion coordinated with four histidines.

The mechanism by which Cu^{2+} ions that are bound to mutant PrP with octarepeat expansions participate in the formation of PrP^{Sc} remains unknown. Whatever the process controlling the formation of prions in the inherited PrP prion diseases, it seems likely to be relevant to understanding other age-dependent neurodegenerative disorders.

That each of the familial and sporadic forms of the neurodegenerative diseases might be caused by prions is consistent with the observation that aging appears to be the most important risk factor. Despite the appeal of prions as an explanation for the late onset of the neurodegenerative diseases, many alternatives have been offered, including age-dependent mitochondrial DNA mutations, oxidative modifications of DNA and proteins, proteasome malfunction, diminished innate immunity, exogenous toxins such as alcohol and drugs, concomitant conditions such as atherosclerosis, somatic mutations, chaperone malfunction, haploinsufficiency, RNA-DNA differences, expanded repeat segments in proteins and noncoding regions of DNA, and postinfectious syndromes (Table 2). Notably, inheritance of the *e4* allele of apolipoprotein E is the only well-established genetic risk factor for sporadic AD.

I posit that precursor proteins become transformed into disease-causing prions through a stochastic process, which most of the time represents a dead-end pathway

Table 2 Some possible explanations for late-onset neurodegeneration

1. Mitochondrial DNA mutations (Coskun et al. 2010; Morais and De Strooper 2010)
2. Oxidative modifications of DNA, lipids, or proteins (Johri and Beal 2012; Larsson 2010)
3. Impaired autophagy (Olanow and McNaught 2006)
4. Altered apoptosis (Yuan and Yankner 2000)
5. Posttranslational chemical modification (Yuan and Yankner 2000)
6. Modified innate immunity (Tracey 2009)
7. Accumulation of exogenous toxins such as heavy metals, alcohol, drugs, and hormones (Chen et al. 2011)
8. Concomitant conditions such as atherosclerosis (Korczyn et al. 2012)
9. RNA-DNA differences (Li et al. 2011)
10. Chaperone malfunction (Macario and Conway de Macario 2005)
11. Somatic mutations (Larsson 2010)
12. Altered regulation of transcription (Bithell et al. 2009)
13. Haploinsufficiency (Deutschbauer et al. 2005)
14. Postinfectious syndromes of the CNS, including late polio, subacute sclerosing leukoencephalitis, postencephalitic Parkinson's disease, and Lyme disease (Greenfield and Matthews 1954; Jubelt 2004; Zilber et al. 1983)
15. Modifier genes like apolipoprotein E and LRRK2 (Castellano et al. 2011; Corti et al. 2011)
16. Polyglutamine expansions (Lee et al. 2012; Wheeler et al. 2002)
17. Noncoding nucleotide repeat expansions (DeJesus-Hernandez et al. 2011; Renton et al. 2011)
18. Cu^{2+} binding to expanded PrP octarepeat region (Stevens et al. 2009)
19. Prion formation and accumulation (Olanow and Prusiner 2009; Prusiner 1989)

where small numbers of prions are cleared via protein degradation pathways. Only when a sufficient number of prions are formed does the titer reach a threshold level that renders the process self-propagating. Under these conditions, prion propagation becomes uncontrolled and eventually CNS dysfunction results.

Fungal Prions

Yeast prions have been invaluable in defining the spectrum of prions (Chien et al. 2004; Halfmann et al. 2012; Wickner 1994). The HET-s prion of *Podospora anserina*, which is a filamentous fungus, controls a programmed cell death reaction termed heterokaryon incompatibility (Greenwald et al. 2010). While yeast prions are not infectious in the sense of being released into the culture medium and infecting other yeast, they are transmissible from mother to daughter cells and, thus, can readily multiply.

Amyloids

The term "amyloid" was initially used to describe waxy substances that accumulate in tissues in a variety of systemic and CNS human disorders. The histological staining properties of amyloid accumulations in tissues resemble those of some long-chain polysaccharides from plants referred to as starch. Some of the most well-studied amyloids include the abnormal immunoglobulins that are produced by neoplastic plasma cells in multiple myeloma. Some of these immunoglobins, like the Bence-Jones light chains, readily polymerize into amyloid fibrils. The accumulation of Bence-Jones paraproteins in the kidney leads to renal failure. Other amyloid proteins such as mutant transthyretin (TTR) show diminished stability as a native tetramer and readily polymerize into amyloid fibrils. TTR amyloid deposition within myelin sheaths and cardiac muscle causes polyneuropathy and cardiomyopathy, respectively. Recently, the drug Tafamidis [2-(3,5-dichloro-phenyl)-benzoxazole-6-carboxylic acid] was developed to stabilize the tetrameric structure of mutant TTR and is now being used to treat familial amyloidotic polyneuropathy (FAP; Johnson et al. 2012). Features common to all amyloid fibrils are their nonbranched ultrastructure, birefringence in polarized light after staining with Congo red dye, shift in the fluorescence emission wavelength after staining with Thioflavin T dye, and their crossed β-pleated sheet structure (Divry 1927; Eisenberg and Jucker 2012; Glenner et al. 1974; Rogers 1965).

In CNS disorders, amyloid plaques are a prominent feature in such diseases as kuru and AD (Alzheimer 1907; Divry 1927; Klatzo et al. 1959). Subsequently, the proteins that polymerize into the fibrils of kuru and AD were identified as PrP^{Sc} and Aβ, respectively (Bolton et al. 1982; Glenner and Wong 1984a). Unlike the Bence-Jones immunoglobulin light chains, both PrP^{Sc} and Aβ are prions in that each can acquire a conformation that becomes self-propagating. While amyloid plaques are an essential feature of AD, they are nonobligatory features of kuru, CJD, scrapie, and other PrP^{Sc} disorders (Prusiner et al. 1990). Like many mammalian prions, most yeast prions have a high β-sheet content and can polymerize into amyloid fibrils.

It is important to distinguish between prions and amyloids (Piro et al. 2011; Wille et al. 2009); as noted above, prions need not polymerize into amyloid fibrils and can undergo self-propagation as oligomers (Alper et al. 1966; Bellinger-Kawahara et al. 1988; McKinley et al. 1986; Silveira et al. 2005). The self-propagation of an alternative conformational state is a key feature of all prions. In contrast, amyloids are linear polymers that increase in number as new seeds are generated. It can be argued that any prion oligomer has the potential to become an amyloid seed; I contend that the polymerization of prions into amyloid fibrils represents a sequestration process whereby the brain seeks to minimize neurotoxicity. While some investigators argue that any protein can be induced to polymerize in an amyloid fibril (Dobson 2003), only a small subset of proteins seems to become prions. Amyloid fibrils are not only pathogenic but they may also act as storage depots for secreted proteins (Goldschmidt et al. 2010).

Synuclein Prions

In the late 1990s, fetal brain cells taken from the substantia nigra of aborted fetuses were transplanted into patients with advanced Parkinson's disease. A decade later, Lewy bodies were found in the grafted fetal brain cells when the Parkinson's patients died (Kordower et al. 2008; Li et al. 2008). The surface of Lewy bodies is covered with fibrils composed of β-sheet-rich, α-synuclein proteins. The normal form of α-synuclein seems to be either unstructured or high in α-helical structure, but like other prion proteins, α-synuclein can adopt a β-sheet–rich conformation. Though unproven, it seems likely that β-sheet–rich α-synuclein prions crossed from the transplanted patient's own neurons into the grafted cells and induced a change in the structure of α-synuclein (Olanow and Prusiner 2009). Once established, this process became self-propagating, as is the case for all pathogenic prions. This scenario is consistent with the findings of Braak and coworkers, who have mapped the spread of aggregated α-synuclein called "Lewy neurites" from the gut, into the brainstem, and throughout the cerebral hemispheres (Braak et al. 1996, 2003; Polymenidou and Cleveland 2012).

Tg mice expressing mutant human α-synuclein(A53T) were found to exhibit neurological disease at ~400 days of age (Giasson et al. 2002). When brains from these ill mice were homogenized and injected into weanlings, the recipient mice became ill at ~200 days of age (Luk et al. 2012; Mougenot et al. 2012). Recombinant α-synuclein was purified and induced to form fibrillar aggregates in vitro. After introduction of the aggregated α-synuclein into cultured cells, the aggregates of α-synuclein underwent self-propagation and therefore became prions (Volpicelli-Daley et al. 2011).

SOD1 Prions

Studies of the progressive spread of motor neuron lesions along the neuraxis suggest an orderly and active process in ALS, also known as Lou Gehrig's disease (Ravits and La Spada 2009). Aggregates of mutant human superoxide dismutase (SOD1) protein have been used to initiate self-propagation in cultured cells, which can continue indefinitely; thus, mutant SOD1(H46R) forms prions (Münch et al. 2011). In another study, two human SOD1s harboring either G127X or G85R induced wt SOD1 to misfold and aggregate in human neural cell lines (Grad et al. 2011). More than 150 different mutations in SOD1 have been found to cause familial forms of ALS (Valentine et al. 2005).

In addition to mutant SOD1, mutations in two RNA-binding proteins, TDP-43 and FUS, have been identified in patients with familial ALS. As well as being involved in RNA metabolism, TDP-43 and FUS form aggregates in neurons in some cases of ALS and FTD (Polymenidou and Cleveland 2012; Udan and Baloh 2011). Recent evidence indicates that both TDP-43 and FUS contain fungal prion-like domains rich in glutamine and asparagine residues, and, in the case of TDP-43, this domain contains a substantial number of disease-causing mutations.

Huntingtin Prions

The wt huntingtin protein harbors an N-terminal region of ~35 glutamine residues. An additional 5–20 glutamines are found in most Huntington's patients but many more have also been recorded (Lee et al. 2012). The length of the polyglutamine expansion is inversely proportional to the age of onset of Huntington's disease. Expanded polyglutamine repeats in a fragment of the huntingtin protein show spontaneous aggregation that self-propagates in cultured cells; in other words, they are prions (Ren et al. 2009). Huntingtin prions explain why people with 5–10 additional glutamines do not become ill until they are 40–50 years of age even though the mutant protein is produced beginning in embryogenesis.

Nonpathogenic Mammalian Prions

Importantly, some prions in mammals such as cytoplasmic polyadenylation element binding protein (CPEB) and mitochondrial antiviral-signaling protein (MAVS) are not known to cause disease but perform important cellular functions (Hou et al. 2011; Si et al. 2010). Both CPEB and MAVS contain glutamine-rich domains like yeast prions. The CPEB prion seems to control localized gene transcription in long-term memory whereas the MAVS prion features in the immune response to infection by some RNA viruses. Unexpectedly, the biologically active forms of CPEB and MAVS are the oligomeric prion states and not the monomeric precursor proteins. Another nonpathogenic prion may be the T-cell–restricted intracellular antigen 1 (TIA-1), which is an RNA-binding protein that promotes translational arrest and the assembly of stress granules under conditions of cellular stress, including acute energy starvation (Gilks et al. 2004). TIA-1 contains a carboxy-terminal, glutamine-rich, prion-related domain that can undergo aggregation and thus functions as a metabolic switch. When this glutamine-rich domain of TIA-1 aggregates, stress granule formation ceases, but the process appears to be reversible.

Concluding Remarks

The convergence of studies of these common neurodegenerative maladies has been remarkable (Table 1). While many mysteries are now explicable within the framework of the prion concept, the science of prions is still in its infancy. Many unanticipated discoveries seem likely to emerge from the future study of prions.

Most important, strategies for developing informative molecular diagnostics and effective therapeutics for these elusive neurodegenerative disorders emerge from our growing knowledge of prions. Early diagnosis will require reporters such as PET ligands to identify prions long before symptoms appear. Meaningful treatments are likely to require cocktails of drugs that diminish the precursor protein, interfere with the conversion of precursors into prions, and/or enhance the clearance of prions.

References

Alper T, Haig DA, Clarke MC (1966) The exceptionally small size of the scrapie agent. Biochem Biophys Res Commun 22:278–284
Alzheimer A (1907) Ueber eine eigenartige erkrankung der hirnrinde. Cent Nervenheilk Psychiat 30:177–179
Baker HF, Ridley RM, Duchen LW, Crow TJ, Bruton CJ (1994) Induction of β(a4)-amyloid in primates by injection of Alzheimer's disease brain homogenate. Mol Neurobiol 8:25–39
Bellinger-Kawahara CG, Kempner E, Groth DF, Gabizon R, Prusiner SB (1988) Scrapie prion liposomes and rods exhibit target sizes of 55,000 Da. Virology 164:537–541
Bithell A, Johnson R, Buckley NJ (2009) Transcriptional dysregulation of coding and non-coding genes in cellular models of Huntington's disease. Biochem Soc Trans 37:1270–1275
Bolton DC, McKinley MP, Prusiner SB (1982) Identification of a protein that purifies with the scrapie prion. Science 218:1309–1311
Braak H, Braak E (1995) Staging of Alzheimer's disease-related neurofibrillary changes. Neurobiol Aging 16:271–284
Braak H, Del Tredici K (2011) Alzheimer's pathogenesis: is there neuron-to-neuron propagation? Acta Neuropathol 121:589–595
Braak H, Braak E, Yilmazer D, de Vos RA, Jansen EN, Bohl J (1996) Pattern of brain destruction in Parkinson's and Alzheimer's diseases. J Neural Transm 103:455–490
Braak H, Rub U, Gai WP, Del Tredici K (2003) Idiopathic Parkinson's disease: possible routes by which vulnerable neuronal types may be subject to neuroinvasion by an unknown pathogen. J Neural Transm 110:517–536
Brion J-P, Passareiro H, Nunez J, Flament-Durand J (1985) Mise en évidence immunologique de la protéine tau au niveau des lésions de dégénérescence neurofibrillaire de la maladie d'Alzheimer. Arch Biol 95:229–235
Castellano JM, Kim J, Stewart FR, Jiang H, DeMattos RB, Patterson BW, Fagan AM, Morris JC, Mawuenyega KG, Cruchaga C, Goate AM, Bales KR, Paul SM, Bateman RJ, Holtzman DM (2011) Human apoE isoforms differentially regulate brain amyloid-beta peptide clearance. Sci Transl Med 3:89ra57
Chen Y-C, Prescott CA, Walsh D, Patterson DG, Riley BP, Kendler KS, Kuo PH (2011) Different phenotypic and genotypic presentations in alcohol dependence: age at onset matters. J Stud Alcohol Drugs 72:752–762
Chien P, Weissman JS, DePace AH (2004) Emerging principles of conformation-based prion inheritance. Annu Rev Biochem 73:617–656
Clavaguera F, Bolmont T, Crowther RA, Abramowski D, Frank S, Probst A, Fraser G, Stalder AK, Beibel M, Staufenbiel M, Jucker M, Goedert M, Tolnay M (2009) Transmission and spreading of tauopathy in transgenic mouse brain. Nat Cell Biol 11:909–913
Corsellis JA, Bruton CJ, Freeman-Browne D (1973) The aftermath of boxing. Psychol Med 3:270–303
Corti O, Lesage S, Brice A (2011) What genetics tells us about the causes and mechanisms of Parkinson's disease. Physiol Rev 91:1161–1218
Coskun PE, Wyrembak J, Derbereva O, Melkonian G, Doran E, Lott IT, Head E, Cotman CW, Wallace DC (2010) Systemic mitochondrial dysfunction and the etiology of Alzheimer's disease and Down syndrome dementia. J Alzheimers Dis 20(Suppl 2):S293–S310
Croes EA, Theuns J, Houwing-Duistermaat JJ, Dermaut B, Sleegers K, Roks G, Van den Broeck M, van Harten B, van Swieten JC, Cruts M, Van Broeckhoven C, van Duijn CM (2004) Octapeptide repeat insertions in the prion protein gene and early onset dementia. J Neurol Neurosurg Psychiatry 75:1166–1170
de Calignon A, Polydoro M, Suarez-Calvet M, William C, Adamowicz DH, Kopeikina KJ, Pitstick R, Sahara N, Ashe KH, Carlson GA, Spires-Jones TL, Hyman BT (2012) Propagation of tau pathology in a model of early Alzheimer's disease. Neuron 73:685–697

DeJesus-Hernandez M, Mackenzie IR, Boeve BF, Boxer AL, Baker M, Rutherford NJ, Nicholson AM, Finch NA, Flynn H, Adamson J, Kouri N, Wojtas A, Sengdy P, Hsiung GY, Karydas A, Seeley WW, Josephs KA, Coppola G, Geschwind DH, Wszolek ZK, Feldman H, Knopman DS, Petersen RC, Miller BL, Dickson DW, Boylan KB, Graff-Radford NR, Rademakers R (2011) Expanded GGGGCC hexanucleotide repeat in noncoding region of C9ORF72 causes chromosome 9p-linked FTD and ALS. Neuron 72:245–256

Deutschbauer AM, Jaramillo DF, Proctor M, Kumm J, Hillenmeyer ME, Davis RW, Nislow C, Giaever G (2005) Mechanisms of haploinsufficiency revealed by genome-wide profiling in yeast. Genetics 169:1915–1925

Divry P (1927) Etude histochimique des plaques seniles. J Belge Neurol Psychiat 27:643–654

Dobson CM (2003) Protein folding and misfolding. Nature 426:884–890

Eisele YS, Bolmont T, Heikenwalder M, Langer F, Jacobson LH, Yan ZX, Roth K, Aguzzi A, Staufenbiel M, Walker LC, Jucker M (2009) Induction of cerebral β-amyloidosis: intracerebral versus systemic Aβ inoculation. Proc Natl Acad Sci USA 106:12926–12931

Eisenberg D, Jucker M (2012) The amyloid state of proteins in human diseases. Cell 148:1188–1203

Frost B, Jacks RL, Diamond MI (2009) Propagation of tau misfolding from the outside to the inside of a cell. J Biol Chem 284:12845–12852

Gajdusek DC (1977) Unconventional viruses and the origin and disappearance of kuru. Science 197:943–960

Games D, Adams D, Alessandrini R, Barbour R, Berthelette P, Blackwell C, Carr T, Clemens J, Donaldson T, Gillespie F, Guido T, Hagopian S, Johnson-Wood K, Khan K, Lee M, Leibowitz P, Lieberburg I, Little S, Masliah E, McConlogue L, Montoya-Zavala M, Mucke L, Paganini L, Penniman E, Power M, Schenk D, Seubert P, Snyder B, Soriano F, Tan H, Vitale J, Wadsworth S, Wolozin B, Zhao J (1995) Alzheimer-type neuropathology in transgenic mice overexpressing V717F β-amyloid precursor protein. Nature 373:523–527

Giasson BI, Duda JE, Quinn SM, Zhang B, Trojanowski JQ, Lee VM (2002) Neuronal α-synucleinopathy with severe movement disorder in mice expressing A53T human α-synuclein. Neuron 34:521–533

Gilks N, Kedersha N, Ayodele M, Shen L, Stoecklin G, Dember LM, Anderson P (2004) Stress granule assembly is mediated by prion-like aggregation of TIA-1. Mol Biol Cell 15:5383–5398

Glenner GG, Wong CW (1984a) Alzheimer's disease: initial report of the purification and characterization of a novel cerebrovascular amyloid protein. Biochem Biophys Res Commun 120:885–890

Glenner GG, Wong CW (1984b) Alzheimer's disease and Down's syndrome: sharing of a unique cerebrovascular amyloid fibril protein. Biochem Biophys Res Commun 122:1131–1135

Glenner GG, Eanes ED, Bladen HA, Linke RP, Termine JD (1974) Beta-pleated sheet fibrils – a comparison of native amyloid with synthetic protein fibrils. J Histochem Cytochem 22:1141–1158

Goate A, Chartier-Harlin M-C, Mullan M, Brown J, Crawford F, Fidani L, Giuffra L, Haynes A, Irving N, James L, Mant R, Newton P, Rooke K, Roques P, Talbot C, Pericak-Vance M, Roses A, Williamson R, Rossor M, Owen M, Hardy J (1991) Segregation of a missense mutation in the amyloid precursor protein gene with familial Alzheimer's disease. Nature 349:704–706

Godec MS, Asher DM, Masters CL, Kozachuk WE, Friedland RP, Gibbs CJ Jr, Gajdusek DC, Rapoport SI, Schapiro MB (1991) Evidence against the transmissibility of Alzheimer's disease. Neurology 41:1320

Goldschmidt L, Teng PK, Riek R, Eisenberg D (2010) Identifying the amylome, proteins capable of forming amyloid-like fibrils. Proc Natl Acad Sci USA 107:3487–3492

Goudsmit J, Morrow CH, Asher DM, Yanagihara RT, Masters CL, Gibbs CJ Jr, Gajdusek DC (1980) Evidence for and against the transmissibility of Alzheimer's disease. Neurology 30:945–950

Grad LI, Guest WC, Yanai A, Pokrishevsky E, O'Neill MA, Gibbs E, Semenchenko V, Yousefi M, Wishart DS, Plotkin SS, Cashman NR (2011) Intermolecular transmission of superoxide dismutase 1 misfolding in living cells. Proc Natl Acad Sci USA 108:16398–16403

Greenfield JG, Matthews WB (1954) Post-encephalitic parkinsonism with amyotrophy. J Neurol Neurosurg Psychiatry 17:50–56

Greenwald J, Buhtz C, Ritter C, Kwiatkowski W, Choe S, Maddelein ML, Ness F, Cescau S, Soragni A, Leitz D, Saupe SJ, Riek R (2010) The mechanism of prion inhibition by HET-S. Mol Cell 38:889–899

Grundke-Iqbal I, Iqbal K, Tung Y-C, Quinlan M, Wisniewski HM, Binder LI (1986) Abnormal phosphorylation of the microtubule-associated protein (tau) in Alzheimer cytoskeletal pathology. Proc Natl Acad Sci USA 83:4913–4917

Guo JL, Lee VM-Y (2011) Seeding of normal tau by pathological tau conformers drives pathogenesis of Alzheimer-like tangles. J Biol Chem 286:15317–15331

Hadlow WJ, Eklund CM, Kennedy RC, Jackson TA, Whitford HW, Boyle CC (1974) Course of experimental scrapie virus infection in the goat. J Infect Dis 129:559–567

Halfmann R, Jarosz DF, Jones SK, Chang A, Lancaster AK, Lindquist S (2012) Prions are a common mechanism for phenotypic inheritance in wild yeasts. Nature 482:363–368

Hardy J, Selkoe DJ (2002) The amyloid hypothesis of Alzheimer's disease: progress and problems on the road to therapeutics. Science 297:353–356

Hong M, Zhukareva V, Vogelsberg-Ragaglia V, Wszolek Z, Reed L, Miller BI, Geschwind DH, Bird TD, McKeel D, Goate A, Morris JC, Wilhelmsen KC, Schellenberg GD, Trojanowksi JQ, Lee VM-Y (1998) Mutation-specific functional impairments in distinct tau isoforms of hereditary FTDP-17. Science 282:1914–1917

Hou F, Sun L, Zheng H, Skaug B, Jiang QX, Chen ZJ (2011) MAVS forms functional prion-like aggregates to activate and propagate antiviral innate immune response. Cell 146:448–461

Hsiao K, Baker HF, Crow TJ, Poulter M, Owen F, Terwilliger JD, Westaway D, Ott J, Prusiner SB (1989) Linkage of a prion protein missense variant to Gerstmann-Sträussler syndrome. Nature 338:342–345

Hsiao K, Chapman P, Nilsen S, Eckman C, Harigaya Y, Younkin S, Yang F, Cole GJ (1996) Correlative memory deficits, $A\beta$ elevation, and amyloid plaques in transgenic mice. Science 274:99–102

Hutton M, Lendon CL, Rizzu P, Baker M, Froelich S, Houlden H, Pickering-Brown S, Chakraverty S, Isaacs A, Grover A, Hackett J, Adamson J, Lincoln S, Dickson D, Davies P, Petersen RC, Stevens M, de Graaff E, Wauters E, van Baren J, Hillebrand M, Joosse M, Kwon JM, Nowotny P, Che LK, Norton J, Morris JC, Reed LA, Trojanowski J, Basun H, Lannfelt L, Neystat M, Fahn S, Dark F, Tannenberg T, Dodd PR, Hayward N, Kwok JBJ, Schofield PR, Andreadis A, Snowden J, Craufurd D, Neary D, Owen F, Oostra BA, Hardy J, Goate A, van Swieten J, Mann D, Lynch T, Heutink P (1998) Association of missense and 5′-splice-site mutations in tau with the inherited dementia FTDP-17. Nature 393:702–705

Johnson SM, Connelly S, Fearns C, Powers ET, Kelly JW (2012) The transthyretin amyloidoses: from delineating the molecular mechanism of aggregation linked to pathology to a regulatory-agency-approved drug. J Mol Biol. doi:10.1016/j.jmb.2011.12.060

Johri A, Beal MF (2012) Antioxidants in Huntington's disease. Biochim Biophys Acta 1822:664–674

Jubelt B (2004) Post-polio syndrome. Curr Treat Options Neurol 6:87–93

Kimberlin RH, Walker CA (1979) Pathogenesis of mouse scrapie: dynamics of agent replication in spleen, spinal cord and brain after infection by different routes. J Comp Pathol 89:551–562

Klatzo I, Gajdusek DC, Zigas V (1959) Pathology of kuru. Lab Invest 8:799–847

Korczyn AD, Vakhapova V, Grinberg LT (2012) Vascular dementia. J Neurol Sci 322(1–2):2–10. doi:10.1016/j.jns.2012.03.027

Kordower JH, Chu Y, Hauser RA, Freeman TB, Olanow CW (2008) Lewy body-like pathology in long-term embryonic nigral transplants in Parkinson's disease. Nat Med 14:504–506

Kosik KS, Joachim CL, Selkoe DJ (1986) Microtubule-associated protein tau is a major antigenic component of paired helical filaments in Alzheimer disease. Proc Natl Acad Sci USA 83:4044–4048

Larsson N-G (2010) Somatic mitochondrial DNA mutations in mammalian aging. Annu Rev Biochem 79:683–706

Lee JM, Ramos EM, Lee JH, Gillis T, Mysore JS, Hayden MR, Warby SC, Morrison P, Nance M, Ross CA, Margolis RL, Squitieri F, Orobello S, Di Donato S, Gomez-Tortosa E, Ayuso C, Suchowersky O, Trent RJ, McCusker E, Novelletto A, Frontali M, Jones R, Ashizawa T, Frank S, Saint-Hilaire MH, Hersch SM, Rosas HD, Lucente D, Harrison MB, Zanko A, Abramson RK, Marder K, Sequeiros J, Paulsen JS, Landwehrmeyer GB, Myers RH, Macdonald ME, Gusella JF (2012) CAG repeat expansion in Huntington disease determines age at onset in a fully dominant fashion. Neurology 78:690–695

Li JY, Englund E, Holton JL, Soulet D, Hagell P, Lees AJ, Lashley T, Quinn NP, Rehncrona S, Bjorklund A, Widner H, Revesz T, Lindvall O, Brundin P (2008) Lewy bodies in grafted neurons in subjects with Parkinson's disease suggest host-to-graft disease propagation. Nat Med 14:501–503

Li M, Wang IX, Li Y, Bruzel A, Richards AL, Toung JM, Cheung VG (2011) Widespread RNA and DNA sequence differences in the human transcriptome. Science 333:53–58

Liu L, Drouet V, Wu JW, Witter MP, Small SA, Clelland C, Duff K (2012) Trans-synaptic spread of tau pathology in vivo. PLoS One 7:e31302

Luk KC, Kehm VM, Zhang B, O'Brien P, Trojanowski JQ, Lee VMY (2012) Intracerebral inoculation of pathological alpha-synuclein initiates a rapidly progressive neurodegenerative alpha-synucleinopathy in mice. J Exp Med 209:975–986

Macario AJL, Conway de Macario E (2005) Sick chaperones, cellular stress, and disease. N Engl J Med 353:1489–1501

Masters CL, Gajdusek DC, Gibbs CJ Jr (1981) Creutzfeldt-Jakob disease virus isolations from the Gerstmann-Sträussler syndrome. Brain 104:559–588

Masters CL, Simms G, Weinman NA, Multhaup G, McDonald BL, Beyreuther K (1985) Amyloid plaque core protein in Alzheimer disease and Down syndrome. Proc Natl Acad Sci USA 82:4245–4249

McKee AC, Cantu RC, Nowinski CJ, Hedley-Whyte ET, Gavett BE, Budson AE, Santini VE, Lee HS, Kubilus CA, Stern RA (2009) Chronic traumatic encephalopathy in athletes: progressive tauopathy after repetitive head injury. J Neuropathol Exp Neurol 68:709–735

McKinley MP, Bolton DC, Prusiner SB (1983) A protease-resistant protein is a structural component of the scrapie prion. Cell 35:57–62

McKinley MP, Braunfeld MB, Bellinger CG, Prusiner SB (1986) Molecular characteristics of prion rods purified from scrapie-infected hamster brains. J Infect Dis 154:110–120

Mead S (2006) Prion disease genetics. Eur J Hum Genet 14:273–281

Meyer-Luehmann M, Coomaraswamy J, Bolmont T, Kaeser S, Schaefer C, Kilger E, Neuenschwander A, Abramowski D, Frey P, Jaton AL, Vigouret JM, Paganetti P, Walsh DM, Mathews PM, Ghiso J, Staufenbiel M, Walker LC, Jucker M (2006) Exogenous induction of cerebral beta-amyloidogenesis is governed by agent and host. Science 313:1781–1784

Morais VA, De Strooper B (2010) Mitochondria dysfunction and neurodegenerative disorders: cause or consequence. J Alzheimers Dis 20(Suppl 2):S255–S263

Morales R, Duran-Aniotz C, Castilla J, Estrada LD, Soto C (2012) De novo induction of amyloid-β deposition in vivo. Mol Psychiatry 17(12):1347–1353. doi:10.1038/mp.2011.120

Mougenot AL, Nicot S, Bencsik A, Morignat E, Verchère J, Lakhdar L, Legastelois S, Baron T (2012) Prion-like acceleration of a synucleinopathy in a transgenic mouse model. Neurobiol Aging 33(9):2225–2228. doi:10.1016/j.neurobiolaging.2011.06.022

Münch C, O'Brien J, Bertolotti A (2011) Prion-like propagation of mutant superoxide dismutase-1 misfolding in neuronal cells. Proc Natl Acad Sci USA 108:3548–3553

Olanow CW, McNaught KS (2006) Ubiquitin-proteasome system and Parkinson's disease. Mov Disord 21:1806–1823

Olanow CW, Prusiner SB (2009) Is Parkinson's disease a prion disorder? Proc Natl Acad Sci USA 106:12571–12572

Omalu BI, DeKosky ST, Minster RL, Kamboh MI, Hamilton RL, Wecht CH (2005) Chronic traumatic encephalopathy in a National Football League player. Neurosurgery 57:128–134

Omalu B, Hammers JL, Bailes J, Hamilton RL, Kamboh MI, Webster G, Fitzsimmons RP (2011) Chronic traumatic encephalopathy in an Iraqi war veteran with posttraumatic stress disorder who committed suicide. Neurosurg Focus 31:E3

Piro JR, Wang F, Walsh DJ, Rees JR, Ma J, Supattapone S (2011) Seeding specificity and ultrastructural characteristics of infectious recombinant prions. Biochemistry 50:7111–7116

Polymenidou M, Cleveland DW (2012) Prion-like spread of protein aggregates in neurodegeneration. J Exp Med 209:889–893

Polymeropoulos MH, Lavedan C, Leroy E, Ide SE, Dehejia A, Dutra A, Pike B, Root H, Rubenstein J, Boyer R, Stenroos ES, Chandrasekharappa S, Athanassiadou A, Papapetropoulos T, Johnson WG, Lazzarini AM, Duvoisin RC, Di Iorio G, Golbe LI, Nussbaum RL (1997) Mutation in the alpha-synuclein gene identified in families with Parkinson's disease. Science 276:2045–2047

Prusiner SB (1982) Novel proteinaceous infectious particles cause scrapie. Science 21:136–144

Prusiner SB (1984) Some speculations about prions, amyloid, and Alzheimer's disease. N Engl J Med 310:661–663

Prusiner SB (1989) Scrapie prions. Annu Rev Microbiol 43:345–374

Prusiner SB (2001) Shattuck lecture – neurodegenerative diseases and prions. N Engl J Med 344:1516–1526

Prusiner SB, McKinley MP, Bowman KA, Bolton DC, Bendheim PE, Groth DF, Glenner GG (1983) Scrapie prions aggregate to form amyloid-like birefringent rods. Cell 35:349–358

Prusiner SB, Scott M, Foster D, Pan K-M, Groth D, Mirenda C, Torchia M, Yang S-L, Serban D, Carlson GA, Hoppe PC, Westaway D, DeArmond SJ (1990) Transgenetic studies implicate interactions between homologous PrP isoforms in scrapie prion replication. Cell 63:673–686

Rabinovici GD, Miller BL (2010) Frontotemporal lobar degeneration: epidemiology, pathophysiology, diagnosis and management. CNS Drugs 24:375–398

Rademakers R, Hutton M (2007) The genetics of frontotemporal lobar degeneration. Curr Neurol Neurosci Rep 7:434–442

Ravits JM, La Spada AR (2009) ALS motor phenotype heterogeneity, focality, and spread: deconstructing motor neuron degeneration. Neurology 73:805–811

Ren PH, Lauckner JE, Kachirskaia I, Heuser JE, Melki R, Kopito RR (2009) Cytoplasmic penetration and persistent infection of mammalian cells by polyglutamine aggregates. Nat Cell Biol 11:219–225

Renton AE, Majounie E, Waite A, Simon-Sanchez J, Rollinson S, Gibbs JR, Schymick JC, Laaksovirta H, van Swieten JC, Myllykangas L, Kalimo H, Paetau A, Abramzon Y, Remes AM, Kaganovich A, Scholz SW, Duckworth J, Ding J, Harmer DW, Hernandez DG, Johnson JO, Mok K, Ryten M, Trabzuni D, Guerreiro RJ, Orrell RW, Neal J, Murray A, Pearson J, Jansen IE, Sondervan D, Seelaar H, Blake D, Young K, Halliwell N, Callister JB, Toulson G, Richardson A, Gerhard A, Snowden J, Mann D, Neary D, Nalls MA, Peuralinna T, Jansson L, Isoviita VM, Kaivorinne AL, Holtta-Vuori M, Ikonen E, Sulkava R, Benatar M, Wuu J, Chio A, Restagno G, Borghero G, Sabatelli M, Heckerman D, Rogaeva E, Zinman L, Rothstein JD, Sendtner M, Drepper C, Eichler EE, Alkan C, Abdullaev Z, Pack SD, Dutra A, Pak E, Hardy J, Singleton A, Williams NM, Heutink P, Pickering-Brown S, Morris HR, Tienari PJ, Traynor BJ (2011) A hexanucleotide repeat expansion in C9ORF72 is the cause of chromosome 9p21-linked ALS-FTD. Neuron 72:257–268

Ridley RM, Baker HF, Windle CP, Cummings RM (2006) Very long term studies of the seeding of beta-amyloidosis in primates. J Neural Transm 113:1243–1251

Roberson ED, Scearce-Levie K, Palop JJ, Yan F, Cheng IH, Wu T, Gerstein H, Yu GQ, Mucke L (2007) Reducing endogenous tau ameliorates amyloid beta-induced deficits in an Alzheimer's disease mouse model. Science 316:750–754

Rogers DR (1965) Screening for amyloid with the thioflavin-T fluorescent method. Am J Clin Pathol 44:59–61

Roos R, Gajdusek DC, Gibbs CJ Jr (1973) The clinical characteristics of transmissible Creutzfeldt-Jakob disease. Brain 96:1–20

Rosen RF, Fritz JJ, Dooyema J, Cintron AF, Hamaguchi T, Lah JJ, Levine H 3rd, Jucker M, Walker LC (2012) Exogenous seeding of cerebral beta-amyloid deposition in beta APP-transgenic rats. J Neurochem 120:660–666

Safar JG, Kellings K, Serban A, Groth D, Cleaver JE, Prusiner SB, Riesner D (2005) Search for a prion-specific nucleic acid. J Virol 79:10796–10806

Schellenberg GD, Bird TD, Wijsman EM, Orr HT, Anderson L, Nemens E, White JA, Bonnycastle L, Weber JL, Alonso ME, Potter H, Heston LL, Martin GM (1992) Genetic linkage evidence for a familial Alzheimer's disease locus on chromosome 14. Science 258:668–671

Scott JR, Davies D, Fraser H (1992) Scrapie in the central nervous system: neuroanatomical spread of infection and Sinc control of pathogenesis. J Gen Virol 73(Pt 7):1637–1644

Seeley WW, Crawford RK, Zhou J, Miller BL, Greicius MD (2009) Neurodegenerative diseases target large-scale human brain networks. Neuron 62:42–52

Si K, Choi YB, White-Grindley E, Majumdar A, Kandel ER (2010) Aplysia CPEB can form prion-like multimers in sensory neurons that contribute to long-term facilitation. Cell 140:421–435

Silveira JR, Raymond GJ, Hughson AG, Race RE, Sim VL, Hayes SF, Caughey B (2005) The most infectious prion protein particles. Nature 437:257–261

Spillantini MG, Schmidt ML, Lee VM-Y, Trojanowski JQ, Jakes R, Goedert M (1997) α-Synuclein in Lewy bodies. Nature 388:839–840

St. George-Hyslop PH (1999) Molecular genetics of Alzheimer disease. In: Terry RD, Katzman R, Bick KL, Sisodia SS (eds) Alzheimer disease, 2nd edn. Lippincott Williams & Wilkins, Philadelphia, pp 311–326

St. George-Hyslop P, Haines J, Rogaev E, Mortilla M, Vaula G, Pericak-Vance M, Foncin J-F, Montesi M, Bruni A, Sorbi S, Rainero I, Pinessi L, Pollen D, Polinsky R, Nee L, Kennedy J, Macciardi F, Rogaeva E, Liang Y, Alexandrova N, Lukiw W, Schlumpf K, Tanzi R, Tsuda T, Farrer L, Cantu J-M, Duara R, Amaducci L, Bergamini L, Gusella J, Roses A, McLachlan DC (1992) Genetic evidence for a novel familial Alzheimer's disease locus on chromosome 14. Nat Genet 2:330–334

Stevens DJ, Walter ED, Rodriguez A, Draper D, Davies P, Brown DR, Millhauser GL (2009) Early onset prion disease from octarepeat expansion correlates with copper binding properties. PLoS Pathog 5:e1000390

Stöhr J, Watts JC, Mensinger ZL, Oehler A, Grillo SK, DeArmond SJ, Prusiner SB, Giles K (2012) Purified and synthetic Alzheimer's Aβ prions. Proc Natl Acad Sci USA 109(27):11025–11030

Taraboulos A, Jendroska K, Serban D, Yang S-L, DeArmond SJ, Prusiner SB (1992) Regional mapping of prion proteins in brains. Proc Natl Acad Sci USA 89:7620–7624

Tatzelt J, Groth DF, Torchia M, Prusiner SB, DeArmond SJ (1999) Kinetics of prion protein accumulation in the CNS of mice with experimental scrapie. J Neuropathol Exp Neurol 58:1244–1249

Tracey KJ (2009) Reflex control of immunity. Nat Rev Immunol 9:418–428

Udan M, Baloh RH (2011) Implications of the prion-related Q/N domains in TDP-43 and FUS. Prion 5:1–5

Valentine JS, Doucette PA, Zittin Potter S (2005) Copper-zinc superoxide dismutase and amyotrophic lateral sclerosis. Annu Rev Biochem 74:563–593

van der Kamp MW, Daggett V (2009) The consequences of pathogenic mutations to the human prion protein. Protein Eng Des Sel 22:461–468

Volpicelli-Daley LA, Luk KC, Patel TP, Tanik SA, Riddle DM, Stieber A, Meaney DF, Trojanowski JQ, Lee VM (2011) Exogenous alpha-synuclein fibrils induce Lewy body pathology leading to synaptic dysfunction and neuron death. Neuron 72:57–71

Watts JC, Giles K, Grillo SK, Lemus A, DeArmond SJ, Prusiner SB (2011) Bioluminescence imaging of Abeta deposition in bigenic mouse models of Alzheimer's disease. Proc Natl Acad Sci USA 108:2528–2533

Wheeler VC, Gutekunst CA, Vrbanac V, Lebel LA, Schilling G, Hersch S, Friedlander RM, Gusella JF, Vonsattel JP, Borchelt DR, MacDonald ME (2002) Early phenotypes that presage late-onset neurodegenerative disease allow testing of modifiers in Hdh CAG knock-in mice. Hum Mol Genet 11:633–640

Wickner RB (1994) [URE3] as an altered URE2 protein: evidence for a prion analog in *Saccharomyces cerevisiae*. Science 264:566–569

Wille H, Bian W, McDonald M, Kendall A, Colby DW, Bloch L, Ollesch J, Boronvinskiy AL, Cohen FE, Prusiner SB, Stubbs G (2009) Natural and synthetic prion structure from X-ray fiber diffraction. Proc Natl Acad Sci USA 106:16990–16995

Wood JG, Mirra SS, Pollock NJ, Binder II (1986) Neurofibrillary tangles of Alzheimer's disease share antigenic determinants with the axonal microtubule-associated protein tau. Proc Natl Acad Sci USA 83:4040–4043

Yuan J, Yankner BA (2000) Apoptosis in the nervous system. Nature 407:802–809

Zilber N, Rannon L, Alter M, Kahana E (1983) Measles, measles vaccination, and risk of subacute sclerosing panencephalitis (SSPE). Neurology 33:1558–1564

β-Amyloid Fibril Structures, In Vitro and In Vivo

Robert Tycko

Abstract Since 1998, a great deal of progress has been made towards determining and understanding the molecular structures of amyloid fibrils, including fibrils formed by the β-amyloid peptide that is associated with Alzheimer's disease. Much of this progress has resulted from solid state nuclear magnetic resonance (NMR) measurements, which provide experimental constraints on molecular conformations and interatomic distances without requiring solubility or crystallinity. In general, amyloid fibrils are polymorphic, meaning that fibrils formed by a given peptide or protein can have multiple, distinct molecular structures, depending on the precise conditions under which the fibrils grow. From solid state NMR, electron microscopy, and other measurements, we have developed two detailed molecular structural models for fibrils formed by the 40-residue wild-type β-amyloid ($A\beta_{1-40}$) peptide. These two $A\beta_{1-40}$ fibril polymorphs share a common, parallel β-sheet organization and contain similar peptide conformations but differ in overall symmetry and in other structural aspects. We have also identified and characterized a surprising antiparallel β-sheet structure in metastable fibrils formed by a disease-associated mutant, D23N-$A\beta_{1-40}$, which reveals how similar sets of interactions can stabilize both parallel and antiparallel β-sheets within amyloid fibrils. We are currently extending our structural studies to β-amyloid fibrils that develop in human brain tissue, with the goal of testing whether variations in fibril structure correlate with variations in severity, progression rate, or other characteristics of Alzheimer's disease.

R. Tycko (✉)
Laboratory of Chemical Physics, National Institute of Diabetes and Digestive and Kidney Diseases, National Institutes of Health, Building 5, Room 112, Bethesda, MD 20892-0520, USA
e-mail: robertty@mail.nih.gov

M. Jucker and Y. Christen (eds.), *Proteopathic Seeds and Neurodegenerative Diseases*, 19
Research and Perspectives in Alzheimer's Disease,
DOI 10.1007/978-3-642-35491-5_2, © Springer-Verlag Berlin Heidelberg 2013

Introduction

Knowledge about the structures of amyloid fibrils is important for several reasons: (1) detailed structural information at the molecular and atomic level is necessary for rational design of compounds that inhibit amyloid formation (Gordon et al. 2001; Sato et al. 2006; Sievers et al. 2011) or bind specifically to amyloid (Klunk et al. 2004; Schutz et al. 2011; Wong et al. 2010). Such compounds have therapeutic and diagnostic applications; (2) substantial evidence exists that amyloid fibrils are polymorphic at the molecular structural level, i.e., that the amino acid sequence alone does not determine the molecular structure uniquely (Goldsbury et al. 2000; Kodali et al. 2010; Luca et al. 2007; Paravastu et al. 2008; Petkova et al. 2005). The possibility exists that fibrils with distinct structures may have distinct biological effects (Meyer-Luehmann et al. 2006; Paravastu et al. 2009; Petkova et al. 2005; Tycko et al. 2009). Details of the molecular structures may therefore have biomedical consequences; and (3) detailed structural information is a requirement for any understanding of the intermolecular forces that drive amyloid formation and for any understanding of the mechanisms and pathways by which amyloid fibrils form from monomeric peptides and proteins (Fawzi et al. 2007; Klimov and Thirumalai 2003).

Research on amyloid fibril structures began in my laboratory in 1998. At that time, relatively little was known about the molecular structures of amyloid fibrils, primarily because amyloid fibrils are inherently noncrystalline and insoluble, making direct structure determination by X-ray crystallography and liquid state nuclear magnetic resonance (NMR) impossible. The only aspect of amyloid fibril structures that was firmly established by experimental data was the fact that amyloid fibrils contain ribbon-like β-sheets, running the length of the fibrils and arranged in a "cross-β" motif, i.e., a motif in which β-strand segments are oriented approximately perpendicular to the long axis of the fibril and are connected by backbone hydrogen bonds that are oriented approximately parallel to the long axis (Sunde et al. 1997). Other structural aspects, including the nature of the β-sheets within the cross-β motif (parallel or antiparallel), the identities of β-strand segments, and even the extent to which amyloid fibrils contain well-defined molecular structures (rather than being highly disordered at the molecular level) were uncertain.

Our understanding of amyloid fibril structures has advanced quite substantially over the past 14 years, to the point where many of the generic features of these structures have been elucidated and detailed structural models, based on large sets of experimental data, have been developed for fibrils formed by specific peptides and proteins (Jaroniec et al. 2004; Paravastu et al. 2008; Petkova et al. 2006; Qiang et al. 2012; Van Melckebeke et al. 2010). The remainder of this article reviews contributions from my laboratory, especially our results for fibrils formed by the β-amyloid (Aβ) peptide that is associated with Alzheimer's disease (AD). Structural models for two distinct polymorphs of Aβ fibrils (Paravastu et al. 2008; Petkova et al. 2006), discussed in detail below, are shown in Fig. 1. Our work relies heavily on solid state NMR methods, which are specialized NMR techniques that are applicable to noncrystalline, insoluble biomolecular systems such as amyloid fibrils. Much of the

Fig. 1 Molecular structural models for Aβ$_{1-40}$ fibrils that were prepared in vitro from synthetic peptide. Cross-sections of the models are shown, with the fibril growth axis perpendicular to the page and with six repeat units. The models have approximate twofold (**a**) or threefold (**b**) symmetry about the growth axis, and represent regularized averages of atomic coordinates in Protein Data Bank files 2LMN (**a**) and 2LMP (**b**). Residues 9–40 are shown, and sidechains of certain residues are indicated. These models were developed from a combination of solid state NMR and electron microscopy data (Paravastu et al. 2008; Petkova et al. 2006)

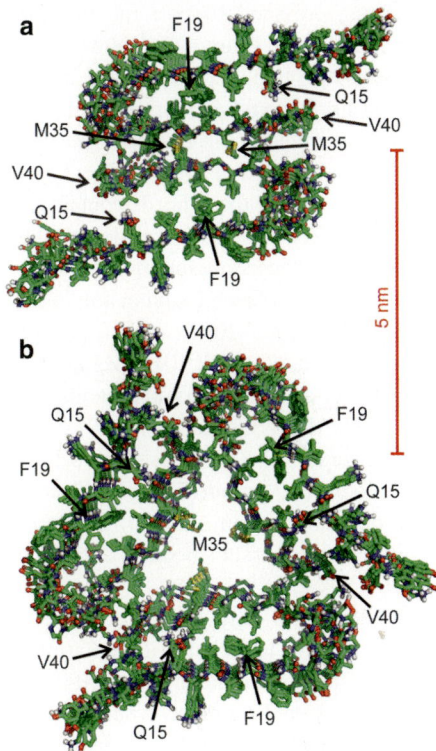

recent progress in our understanding of amyloid fibril structures can be attributed directly to information from solid state NMR measurements (Benzinger et al. 1998; Cheng et al. 2011; Comellas et al. 2011; Debelouchina et al. 2010; Heise et al. 2005; Helmus et al. 2008; Jaroniec et al. 2004; Kammerer et al. 2004; Lansbury et al. 1995; Nielsen et al. 2009; Van Melckebeke et al. 2010). The principles behind these measurements have been reviewed elsewhere (Tycko 2006, 2011). The following discussion focuses on the results we have obtained with solid state NMR methods, supplemented by electron microscopy and other physical measurements.

Organization of β-Sheets in Amyloid Fibrils

The naturally occurring Aβ peptide is primarily 40 or 42 residues in length (Aβ$_{1-40}$ or Aβ$_{1-42}$), with Aβ$_{1-40}$ accounting for roughly 80 % of the Aβ molecules in humans. Early models for fibrils formed by full-length Aβ assumed that the cross-β motif was constructed from antiparallel β-sheets (Chaney et al. 1998; George and Howlett 1999; Lazo and Downing 1998; Li et al. 1999; Tjernberg et al. 1999). Solid state NMR experiments on fibrils formed by a nine-residue peptide, representing residues 34–42 of $A\beta_{1-42}$, provided support for an antiparallel β-sheet structure (Lansbury et al. 1995). However, subsequent solid state NMR experiments on fibrils formed by a 26-residue peptide, representing residues 10–35, provided the first evidence that amyloid fibrils could contain parallel β-sheets, in which neighboring peptide chains are aligned precisely "in-register" (Benzinger et al. 1998). Our own measurements on $A\beta_{1-40}$ (Antzutkin et al. 2000; Balbach et al. 2002) and $A\beta_{1-42}$ (Antzutkin et al. 2002) fibrils showed that these fibrils contain the same in-register, parallel β-sheet organization. Parallel β-sheets in full-length Aβ fibrils were also found by electron paramagnetic resonance measurements (Torok et al. 2002). Moreover, in-register parallel β-sheets have also been found in amyloid fibrils formed by the islet amyloid polypeptide (Luca et al. 2007), tau (Margittai and Langen 2004), $β_2$-microglobulin (Debelouchina et al. 2010), α-synuclein (Der-Sarkissian et al. 2003), and both mammalian and yeast prion proteins (Baxa et al. 2007; Chan et al. 2005; Cobb et al. 2007; Kryndushkin et al. 2011; Shewmaker et al. 2006; Tycko et al. 2010; Wickner et al. 2008). Fibrils formed by the fungal prion protein HET-s have been shown to have a "pseudo-in-register" structure, in which homologous segments align with one another in parallel β-sheets (Van Melckebeke et al. 2010). Thus, the in-register, parallel β-sheet organization predominates in biologically relevant fibrils.

In addition to $A\beta_{34-42}$, several other short Aβ fragments have been shown by solid state NMR to form fibrils containing antiparallel β-sheets (Balbach et al. 2000; Bu et al. 2007; Kammerer et al. 2004; Petkova et al. 2004). These peptides contain only one β-strand segment, whereas full-length Aβ and other full-length amyloid-forming peptides and proteins contain two or more β-strand segments. In addition, both parallel and antiparallel β-sheets have been observed in crystal structures of short amyloid-forming peptides (Sawaya et al. 2007). These observations gave rise to the idea that antiparallel β-sheet structures are limited to fibrils that are formed by short peptides with one β-strand segment and that optimal interactions within and between β-sheets in a cross-β motif generally require a parallel β-sheet organization if more than one β-strand segment is involved.

It was therefore surprising to us when we recently found that fibrils formed by the Asp23-to-Asp mutant of $A\beta_{1-40}$ (D23N-$A\beta_{1-40}$), which produces familial early-onset neurodegeneration (Grabowski et al. 2001), could contain antiparallel β-sheets (Tycko et al. 2009). A molecular model for antiparallel D23N-$A\beta_{1-40}$ fibrils, developed from a substantial set of solid state NMR and electron microscopy measurements (Qiang et al. 2012), is shown in Fig. 2a. Figure 2b shows schematically how parallel and antiparallel β-sheet structures can have similar combinations of favorable hydrophobic interactions, as suggested by our results for D23N-$A\beta_{1-40}$ fibrils. On the other hand, experimental data show conclusively that antiparallel D23N-$A\beta_{1-40}$ structures are only metastable, gradually converting to parallel structures when the two types of structures are mixed (Qiang et al. 2011, 2012).

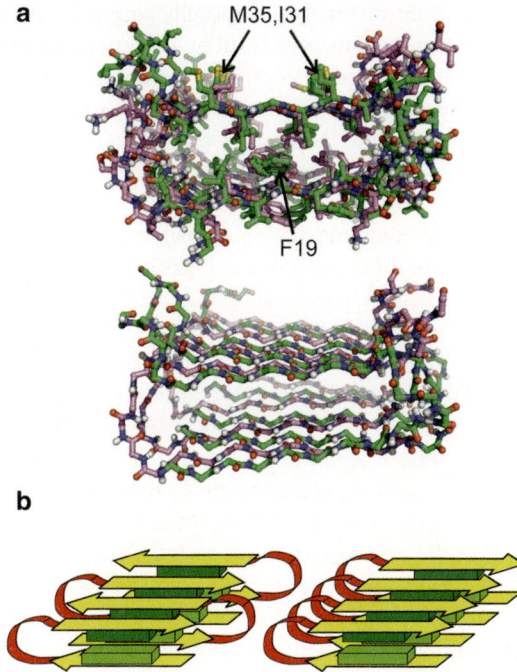

Fig. 2 (**a**) Molecular structural model for D23N-Aβ$_{1-40}$ fibrils that contain antiparallel β-sheets, showing all atoms except non-polar hydrogens (*top*) or only backbone atoms (*bottom*). This model is the first set of atomic coordinates in Protein Data Bank file 2LNQ (Qiang et al. 2012). Residues 15–40 are shown, and the model is viewed in cross-section, with the fibril growth axis nearly perpendicular to the page. Carbon atoms of successive D23N-Aβ$_{1-40}$ molecules are alternately colored *green* or *magenta* to emphasize the antiparallel organization. (**b**) Schematic representation of the antiparallel (*left*) and parallel (*right*) cross-β structures in D23N-Aβ$_{1-40}$ and wild-type Aβ$_{1-40}$ fibrils, showing how both structures produce favorable contacts among sidechains of hydrophobic segments (*green bars*)

Polymorphism of Amyloid Fibrils

One of the central principles of biology and biochemistry is that protein structures are fully and uniquely determined by amino acid sequences. This principle does not apply to amyloid fibrils. We have shown that subtle variations in growth conditions lead to significant, reproducible, and self-propagating variations in the molecular structures of Aβ$_{1-40}$ fibrils (Petkova et al. 2005; Qiang et al. 2011). Structural variations at the molecular level lead to obvious variations in solid state NMR spectra, in addition to variations in the appearance of the fibrils in electron microscope images (i.e., variations in fibril morphology). The same ideas apply to fibrils formed by most other amyloid-forming peptides and proteins, including mammalian and yeast prion proteins, where self-propagating structural variations are almost certainly the source of distinct prion strains or variants (Bessen and Marsh 1992; Telling et al. 1996; Toyama et al. 2007).

Given that amyloid fibrils are highly polymorphic, one must be very careful when comparing structural measurements performed by different research groups on amyloid fibrils formed by the same peptide or protein. Unless the fibrils are known to have been prepared identically and to exhibit precisely the same morphologies (and ideally to exhibit the same solid state NMR spectra), there is no reason to expect measurements by different research groups to be in quantitative, or even qualitative, agreement.

Some of the sources of amyloid polymorphism are revealed by the structural models in Figs. 1 and 2. In these models, the β-sheet-forming segments are nearly the same and the overall peptide conformations are similar. The largest differences are in (1) overall symmetry, with the models for wild-type $A\beta_{1-40}$ fibrils having either twofold or threefold rotational symmetry about the fibril growth direction and the model for $D23N-A\beta_{1-40}$ fibrils having no rotational symmetry and (2) the nature of the β-sheets, as discussed above. The details of the conformations of the non-β-strand segments, the details of contacts between β-sheets, and the fraction of the peptide sequence that becomes structurally ordered in the fibrils (Bertini et al. 2011) also vary. We find that β-sheets formed by the N-terminal and C-terminal β-strands of $A\beta_{1-40}$ are "staggered," meaning that the two β-strands from a given peptide molecule do not make contact with one another (Paravastu et al. 2008; Petkova et al. 2006). The direction of stagger is difficult to determine from experimental measurements and could also be a source of polymorphism.

It is worth noting that fibrils formed by HET-s are not polymorphic, except under extreme pH conditions (Mizuno et al. 2011; Wasmer et al. 2008), that distinct HET-s prion strains have not been observed, that HET-s fibrils are proposed to have an evolved biological function (Coustou et al. 1997), and that solid state NMR spectra of HET-s fibrils are exceptionally well-resolved and reproducible (Tycko and Hu 2010; Van Melckebeke et al. 2010). As mentioned above, the molecular structure of HET-s fibrils is also unusual (Van Melckebeke et al. 2010), suggesting that HET-s fibrils are a rather unique system.

Measurements in cultures of primary embryonic rat hippocampal neurons indicated statistically significant, although not very large, differences in toxicity for $A\beta_{1-40}$ fibrils with different morphologies (Petkova et al. 2005). In these experiments, the fibrils were prepared in vitro from synthetic $A\beta_{1-40}$. These results do not necessarily imply differences in neurotoxicity in the human brain, but such differences certainly seem possible. In the case of transmissible spongiform encephalopathies caused by mammalian PrP, there is strong evidence that distinct molecular structures of PrP aggregates produce distinct strains, characterized by distinct patterns of PrP deposition and incubation periods between infection and neurodegeneration (Bessen and Marsh 1992; Telling et al. 1996). Analogous phenomena may exist in AD. Conflicting observations regarding correlations between cognitive impairment in AD and $A\beta$ amyloid deposition (Aizenstein et al. 2008; Cummings et al. 1996) can be understood if the neurotoxicity of the amyloid deposits varies with the molecular structure of the $A\beta$ fibrils. Studies on transgenic animals have shown that the histopathology of amyloid deposition induced by inoculation with amyloid-containing tissue extracts depends on the source of the inoculums (Meyer-Luehmann et al. 2006). These studies also show that amyloid deposits within brain tissue cannot be induced by synthetic Aβ material, suggesting the possibility of structural differences between purely synthetic fibrils and the fibrils that develop within brain tissue. Thus, the phenomenon of amyloid polymorphism may have consequences for AD and other amyloid diseases.

Structures of Nonfibrillar Aggregates

In electron microscopy or atomic force microscopy studies of amyloid formation in vitro, one commonly observes nonfibrillar aggregated structures in addition to the bona fide, mature fibrils (Goldsbury et al. 2005). These structures, described by various terms that include "oligomer" and "protofibril," generally appear early in the amyloid formation process and eventually disappear as the amyloid-forming peptide or protein converts fully to the fibrillar state. Such nonfibrillar structures have attracted great interest as toxic agents in AD and other amyloid diseases. Due to the transient nature of oligomers and protofibrils, as well as the difficulty of preparing structurally homogeneous samples, the molecular structures within nonfibrillar aggregates are less well known than the structures within certain amyloid fibrils. However, rather surprisingly in light of the large morphological differences between nonfibrillar and fibrillar aggregates, structural information that has been obtained to date suggests that the molecular structures within oligomers and protofibrils closely resemble the structures within mature fibrils. For wild-type Aβ, studies by electron microscopy and by solid state NMR indicate that elements of both the molecular conformation and the β-sheet organization found in mature fibrils are retained in oligomers and protofibrils (Ahmed et al. 2010; Chimon et al. 2007; Goldsbury et al. 2005; Kheterpal et al. 2006). Results for the D23N-Aβ$_{1-40}$ fibrils described above constitute a case in which the molecular conformation is largely retained but the β-sheet organization is quite different in a metastable fibrillar state that resembles protofibrils (Qiang et al. 2012).

Structural Studies of Aβ Fibrils from Human Brain Tissue

To clarify the connections between amyloid structures and amyloid diseases, we obviously need to identify and characterize structures that actually develop in human tissue. Solid state NMR methods that have been used successfully to characterize the molecular structures of amyloid fibrils require milligrams of isotopically (^{13}C and ^{15}N) labeled fibrils. Thus, direct structural measurements on amyloid extracted from tissue are not possible. However, we have shown that amyloid fibrils extracted from brain tissue of AD patients, obtained at autopsy, can be used as a "seed"

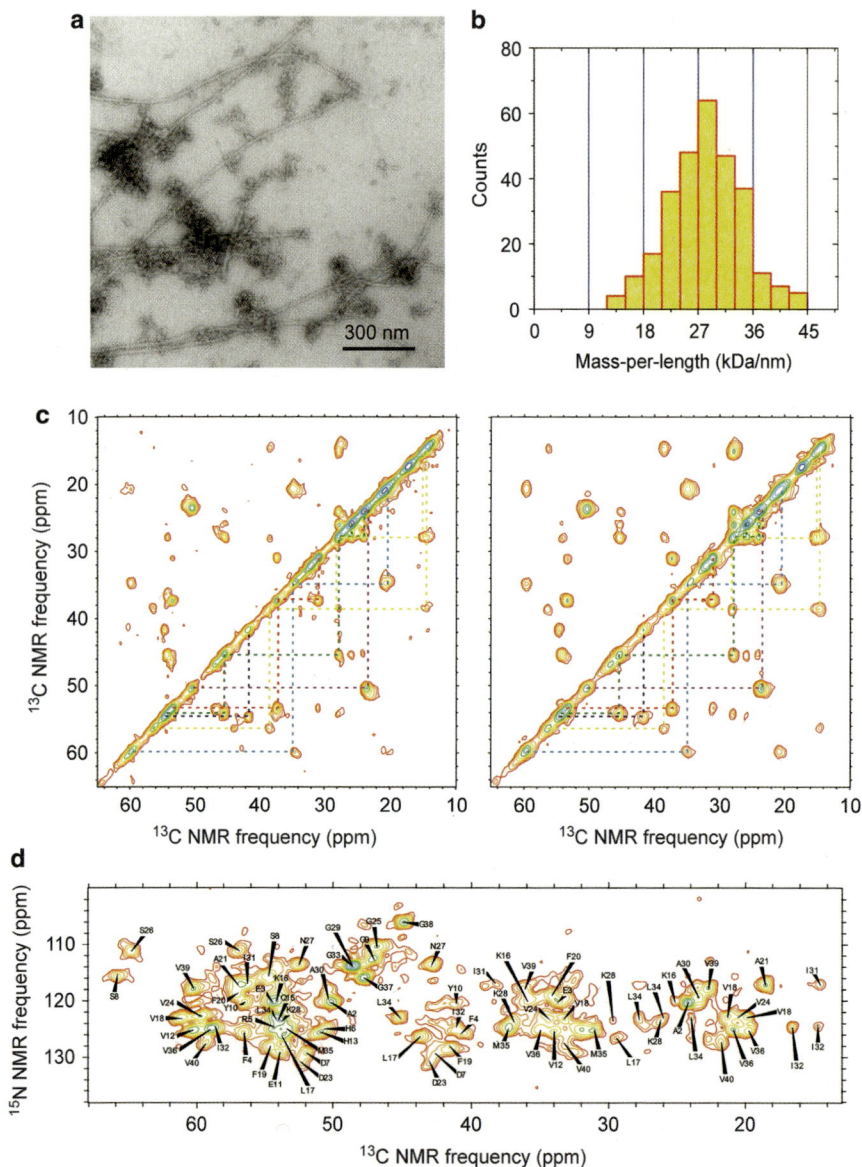

Fig. 3 Structural measurements on brain-derived Aβ$_{1-40}$ fibrils. (**a**) Negatively stained transmission electron microscope image of fibrils prepared by seeding synthetic Aβ$_{1-40}$ with amyloid extracted from brain tissue of an 72-year-old female Alzheimer's disease patient. (**b**) Histogram of mass-per-length values determined from dark-field transmission electron microscope images of unstained fibrils (Chen et al. 2009). Aβ$_{1-40}$ fibrils with twofold and threefold symmetry have values of 18 kDa/nm and 27 kDa/nm, respectively. (**c**) Contour plots of 2D ^{13}C–^{13}C solid state NMR spectra of fibrils derived from temporal/parietal lobe tissue (*left*) and occipital lobe tissue (*right*) of the same patient. *Dashed lines* connect ^{13}C NMR signals from the isotopically labeled

to grow fibrils from synthetic or recombinant $A\beta_{1-40}$, thereby allowing both amplification from the microgram to the milligram scale and isotopic labeling (Paravastu et al. 2009). Studies of seeded fibril growth in vitro have shown that, when fibrils are grown from seeds (i.e., short fibril fragments, generally produced by sonication of longer fibrils), the resulting fibrils retain the molecular structure of the seeds (Petkova et al. 2005), although the relative abundances of different polymorphs within a heterogeneous mixture can evolve as seeded growth is performed through multiple generations (Paravastu et al. 2008; Qiang et al. 2011). Initial experiments with brain tissue used a rigorous extraction procedure to isolate nearly pure $A\beta$ fibrils from the tissue (Paravastu et al. 2009). Three generations of seeded growth were used to create 5- to 10-mg samples, starting with 10 μg of amyloid extract. Solid state NMR spectra indicated a mixture of structures but with the same predominant structure being derived from brain tissue from two different AD patients. Moreover, the NMR chemical shifts in spectra of brain-derived fibrils were different from the chemical shifts in spectra of the synthetic fibrils for which structural models are shown in Fig. 1.

More recently, we have developed a simplified extraction and seeding protocol that requires only 1–2 g of brain tissue, is unlikely to alter fibril structures or structural distributions in the brain tissue, and produces sufficient quantities of isotopically labeled fibrils in a single generation of seeded growth. Figure 3a, b shows electron microscopy data for $A\beta_{1-40}$ fibrils that were grown from brain tissue of an 72-year-old female AD patient. Figure 3c, d shows solid state ^{13}C NMR spectra of these fibrils. A single set of NMR signals appears in these spectra, indicating the predominance of a single fibril structure. Moreover, tissue samples from two brain regions of the same patient (occipital lobe and temporal/parietal lobe) produce the same set of strong NMR signals, indicating the same predominant fibril structure in both regions. These data, plus many additional solid state NMR measurements on fibrils derived from the same tissue samples, imply a molecular structure that is again different from structural models in Fig. 1. Development of a full molecular structural model for the fibrils that developed in the brain of this particular AD patient is currently in progress.

We are now applying the same approach to brain tissue samples from AD patients with diverse clinical histories. The results may reveal whether variations in $A\beta$ fibril structure correlate with variations in AD progression rate, severity, or other characteristics. A similar approach can be applied to other amyloid diseases. If we find that the development of amyloid diseases depends on the details of amyloid fibril structures, it then becomes important to use structural information from solid state NMR, electron microscopy, and other sources to guide the development of diagnostic imaging agents and fibril inhibitors that have precise structural specificity. Thus, we

---

Fig. 3 (continued) residues F19 (*dark blue*), V24 (*light blue*), A30 (*purple*), I31 (*yellow*). L34 (*green*), and M35 (*red*). The same main signals are observed in both spectra. (**d**) 2D ^{15}N–^{13}C solid state NMR spectrum of fibrils derived from occipital lobe tissue, with assignments of the NMR signals to specific residues. Data were acquired by Dr. Junxia Lu

have reached the stage where advanced physical measurement techniques, and the understanding of amyloid structure that has resulted from these techniques, can be applied directly to problems with important consequences for human health.

Acknowledgments This work was supported by the Intramural Research Program of the National Institute of Diabetes and Digestive and Kidney Diseases, a component of the U.S. National Institutes of Health. I thank present and past members of my research group, including Drs. Oleg Antzutkin, Yoshitaka Ishii, John Balbach, Nathan Oyler, Jerry Chan, Aneta Petkova, Anant Paravastu, Kent Thurber, Junxia Lu, and Wei Qiang, for their many contributions to this work. I also thank Prof. Stephen C. Meredith of the University of Chicago for collaborating on several aspects of this work.

References

Ahmed M, Davis J, Aucoin D, Sato T, Ahuja S, Aimoto S, Elliott JI, Van Nostrand WE, Smith SO (2010) Structural conversion of neurotoxic amyloid-β1-42 oligomers to fibrils. Nat Struct Mol Biol 17:561–567

Aizenstein HJ, Nebes RD, Saxton JA, Price JC, Mathis CA, Tsopelas ND, Ziolko SK, James JA, Snitz BE, Houck PR, Bi WZ, Cohen AD, Lopresti BJ, DeKosky ST, Halligan EM, Klunk WE (2008) Frequent amyloid deposition without significant cognitive impairment among the elderly. Arch Neurol 65:1509–1517

Antzutkin ON, Balbach JJ, Leapman RD, Rizzo NW, Reed J, Tycko R (2000) Multiple quantum solid state NMR indicates a parallel, not antiparallel, organization of b-sheets in Alzheimer's β-amyloid fibrils. Proc Natl Acad Sci USA 97:13045–13050

Antzutkin ON, Leapman RD, Balbach JJ, Tycko R (2002) Supramolecular structural constraints on Alzheimer's β-amyloid fibrils from electron microscopy and solid state nuclear magnetic resonance. Biochemistry 41:15436–15450

Balbach JJ, Ishii Y, Antzutkin ON, Leapman RD, Rizzo NW, Dyda F, Reed J, Tycko R (2000) Amyloid fibril formation by Ab16-22, a seven-residue fragment of the Alzheimer's β-amyloid peptide, and structural characterization by solid state NMR. Biochemistry 39:13748–13759

Balbach JJ, Petkova AT, Oyler NA, Antzutkin ON, Gordon DJ, Meredith SC, Tycko R (2002) Supramolecular structure in full-length Alzheimer's b-amyloid fibrils: evidence for a parallel β-sheet organization from solid state nuclear magnetic resonance. Biophys J 83:1205–1216

Baxa U, Wickner RB, Steven AC, Anderson DE, Marekov LN, Yau WM, Tycko R (2007) Characterization of β-sheet structure in Ure2p1-89 yeast prion fibrils by solid state nuclear magnetic resonance. Biochemistry 46:13149–13162

Benzinger TLS, Gregory DM, Burkoth TS, Miller-Auer H, Lynn DG, Botto RE, Meredith SC (1998) Propagating structure of Alzheimer's β-amyloid10-35 is parallel β-sheet with residues in exact register. Proc Natl Acad Sci USA 95:13407–13412

Bertini I, Gonnelli L, Luchinat C, Mao JF, Nesi A (2011) A new structural model of Aβ40 fibrils. J Am Chem Soc 133:16013–16022

Bessen RA, Marsh RF (1992) Biochemical and physical properties of the prion protein from two strains of the transmissible mink encephalopathy agent. J Virol 66:2096–2101

Bu ZM, Shi Y, Callaway DJE, Tycko R (2007) Molecular alignment within b-sheets in Aβ14-23 fibrils: solid state NMR experiments and theoretical predictions. Biophys J 92:594–602

Chan JCC, Oyler NA, Yau WM, Tycko R (2005) Parallel β-sheets and polar zippers in amyloid fibrils formed by residues 10-39 of the yeast prion protein Ure2p. Biochemistry 44:10669–10680

Chaney MO, Webster SD, Kuo YM, Roher AE (1998) Molecular modeling of the Aβ1-42 peptide from Alzheimer's disease. Protein Eng 11:761–767

Chen B, Thurber KR, Shewmaker F, Wickner RB, Tycko R (2009) Measurement of amyloid fibril mass-per-length by tilted-beam transmission electron microscopy. Proc Natl Acad Sci USA 106:14339–14344

Cheng HM, Tsai TWT, Huang WYC, Lee HK, Lian HY, Chou FC, Mou Y, Chan JCC (2011) Steric zipper formed by hydrophobic peptide fragment of Syrian hamster prion protein. Biochemistry 50:6815–6823

Chimon S, Shaibat MA, Jones CR, Calero DC, Aizezi B, Ishii Y (2007) Evidence of fibril-like b-sheet structures in a neurotoxic amyloid intermediate of Alzheimer's β-amyloid. Nat Struct Mol Biol 14:1157–1164

Cobb NJ, Sonnichsen FD, McHaourab H, Surewicz WK (2007) Molecular architecture of human prion protein amyloid: a parallel, in-register β-structure. Proc Natl Acad Sci USA 104:18946–18951

Comellas G, Lemkau LR, Nieuwkoop AJ, Kloepper KD, Ladror DT, Ebisu R, Woods WS, Lipton AS, George JM, Rienstra CM (2011) Structured regions of α-synuclein fibrils include the early-onset Parkinson's disease mutation sites. J Mol Biol 411:881–895

Coustou V, Deleu C, Saupe S, Begueret J (1997) The protein product of the HET-s heterokaryon incompatibility gene of the fungus Podospora anserina behaves as a prion analog. Proc Natl Acad Sci USA 94:9773–9778

Cummings BJ, Pike CJ, Shankle R, Cotman CW (1996) β-Amyloid deposition and other measures of neuropathology predict cognitive status in Alzheimer's disease. Neurobiol Aging 17:921–933

Debelouchina GT, Platt GW, Bayro MJ, Radford SE, Griffin RG (2010) Intermolecular alignment in b2-microglobulin amyloid fibrils. J Am Chem Soc 132:17077–17079

Der-Sarkissian A, Jao CC, Chen J, Langen R (2003) Structural organization of α-synuclein fibrils studied by site-directed spin labeling. J Biol Chem 278:37530–37535

Fawzi NL, Okabe Y, Yap EH, Head-Gordon T (2007) Determining the critical nucleus and mechanism of fibril elongation of the Alzheimer's Aβ1-40 peptide. J Mol Biol 365:535–550

George AR, Howlett DR (1999) Computationally derived structural models of the β-amyloid found in Alzheimer's disease plaques and the interaction with possible aggregation inhibitors. Biopolymers 50:733–741

Goldsbury C, Wirtz S, Muller SA, Sunderji S, Wicki P, Aebi U, Frey P (2000) Studies on the in vitro assembly of Ab1-40: implications for the search for Aβ fibril formation inhibitors. J Struct Biol 130:217–231

Goldsbury C, Frey P, Olivieri V, Aebi U, Muller SA (2005) Multiple assembly pathways underlie amyloid-β fibril polymorphisms. J Mol Biol 352:282–298

Gordon DJ, Sciarretta KL, Meredith SC (2001) Inhibition of β-amyloid(40) fibrillogenesis and disassembly of β-amyloid(40) fibrils by short β-amyloid congeners containing N-methyl amino acids at alternate residues. Biochemistry 40:8237–8245

Grabowski TJ, Cho HS, Vonsattel JPG, Rebeck GW, Greenberg SM (2001) Novel amyloid precursor protein mutation in an Iowa family with dementia and severe cerebral amyloid angiopathy. Ann Neurol 49:697–705

Heise H, Hoyer W, Becker S, Andronesi OC, Riedel D, Baldus M (2005) Molecular-level secondary structure, polymorphism, and dynamics of full-length α-synuclein fibrils studied by solid state NMR. Proc Natl Acad Sci USA 102:15871–15876

Helmus JJ, Surewicz K, Nadaud PS, Surewicz WK, Jaroniec CP (2008) Molecular conformation and dynamics of the Y145Stop variant of human prion protein. Proc Natl Acad Sci USA 105:6284–6289

Jaroniec CP, MacPhee CE, Bajaj VS, McMahon MT, Dobson CM, Griffin RG (2004) High-resolution molecular structure of a peptide in an amyloid fibril determined by magic angle spinning NMR spectroscopy. Proc Natl Acad Sci USA 101:711–716

Kammerer RA, Kostrewa D, Zurdo J, Detken A, Garcia-Echeverria C, Green JD, Muller SA, Meier BH, Winkler FK, Dobson CM, Steinmetz MO (2004) Exploring amyloid formation by a de novo design. Proc Natl Acad Sci USA 101:4435–4440

Kheterpal I, Chen M, Cook KD, Wetzel R (2006) Structural differences in Aβ amyloid protofibrils and fibrils mapped by hydrogen exchange/mass spectrometry with on-line proteolytic fragmentation. J Mol Biol 361:785–795

Klimov DK, Thirumalai D (2003) Dissecting the assembly of Aβ16-22 amyloid peptides into antiparallel β sheets. Structure 11:295–307

Klunk WE, Engler H, Nordberg A, Wang YM, Blomqvist G, Holt DP, Bergstrom M, Savitcheva I, Huang GF, Estrada S, Ausen B, Debnath ML, Barletta J, Price JC, Sandell J, Lopresti BJ, Wall A, Koivisto P, Antoni G, Mathis CA, Langstrom B (2004) Imaging brain amyloid in Alzheimer's disease with Pittsburgh compound B. Ann Neurol 55:306–319

Kodali R, Williams AD, Chemuru S, Wetzel R (2010) Aβ1-40 forms five distinct amyloid structures whose β-sheet contents and fibril stabilities are correlated. J Mol Biol 401:503–517

Kryndushkin DS, Wickner RB, Tycko R (2011) The core of Ure2p prion fibrils is formed by the N-terminal segment in a parallel cross-β structure: evidence from solid state NMR. J Mol Biol 409:263–277

Lansbury PT, Costa PR, Griffiths JM, Simon EJ, Auger M, Halverson KJ, Kocisko DA, Hendsch ZS, Ashburn TT, Spencer RGS, Tidor B, Griffin RG (1995) Structural model for the β-amyloid fibril based on interstrand alignment of an antiparallel sheet comprising a C-terminal peptide. Nat Struct Biol 2:990–998

Lazo ND, Downing DT (1998) Amyloid fibrils may be assembled from b-helical protofibrils. Biochemistry 37:1731–1735

Li LP, Darden TA, Bartolotti L, Kominos D, Pedersen LG (1999) An atomic model for the pleated β-sheet structure of Aβ amyloid protofilaments. Biophys J 76:2871–2878

Luca S, Yau WM, Leapman R, Tycko R (2007) Peptide conformation and supramolecular organization in amylin fibrils: constraints from solid state NMR. Biochemistry 46:13505–13522

Margittai M, Langen R (2004) Template-assisted filament growth by parallel stacking of tau. Proc Natl Acad Sci USA 101:10278–10283

Meyer-Luehmann M, Coomaraswamy J, Bolmont T, Kaeser S, Schaefer C, Kilger E, Neuenschwander A, Abramowski D, Frey P, Jaton AL, Vigouret JM, Paganetti P, Walsh DM, Mathews PM, Ghiso J, Staufenbiel M, Walker LC, Jucker M (2006) Exogenous induction of cerebral β-amyloidogenesis is governed by agent and host. Science 313:1781–1784

Mizuno N, Baxa U, Steven AC (2011) Structural dependence of HET-s amyloid fibril infectivity assessed by cryoelectron microscopy. Proc Natl Acad Sci USA 108:3252–3257

Nielsen JT, Bjerring M, Jeppesen MD, Pedersen RO, Pedersen JM, Hein KL, Vosegaard T, Skrydstrup T, Otzen DE, Nielsen NC (2009) Unique identification of supramolecular structures in amyloid fibrils by solid state NMR spectroscopy. Angew Chem Int Ed Engl 48:2118–2121

Paravastu AK, Leapman RD, Yau WM, Tycko R (2008) Molecular structural basis for polymorphism in Alzheimer's β-amyloid fibrils. Proc Natl Acad Sci USA 105:18349–18354

Paravastu AK, Qahwash I, Leapman RD, Meredith SC, Tycko R (2009) Seeded growth of β-amyloid fibrils from Alzheimer's brain-derived fibrils produces a distinct fibril structure. Proc Natl Acad Sci USA 106:7443–7448

Petkova AT, Buntkowsky G, Dyda F, Leapman RD, Yau WM, Tycko R (2004) Solid state NMR reveals a pH-dependent antiparallel β-sheet registry in fibrils formed by a β-amyloid peptide. J Mol Biol 335:247–260

Petkova AT, Leapman RD, Guo ZH, Yau WM, Mattson MP, Tycko R (2005) Self-propagating, molecular-level polymorphism in Alzheimer's β-amyloid fibrils. Science 307:262–265

Petkova AT, Yau WM, Tycko R (2006) Experimental constraints on quaternary structure in Alzheimer's β-amyloid fibrils. Biochemistry 45:498–512

Qiang W, Yau WM, Tycko R (2011) Structural evolution of Iowa mutant β-amyloid fibrils from polymorphic to homogeneous states under repeated seeded growth. J Am Chem Soc 133:4018–4029

Qiang W, Yau WM, Luo Y, Mattson MP, Tycko R (2012) Antiparallel β-sheet architecture in Iowa-mutant β-amyloid fibrils. Proc Natl Acad Sci USA 109:4443–4448

Sato T, Kienlen-Campard P, Ahmed M, Liu W, Li HL, Elliott JI, Aimoto S, Constantinescu SN, Octave JN, Smith SO (2006) Inhibitors of amyloid toxicity based on β-sheet packing of Aβ 40 and Aβ 42. Biochemistry 45:5503–5516

Sawaya MR, Sambashivan S, Nelson R, Ivanova MI, Sievers SA, Apostol MI, Thompson MJ, Balbirnie M, Wiltzius JJW, McFarlane HT, Madsen AO, Riekel C, Eisenberg D (2007) Atomic structures of amyloid cross-β spines reveal varied steric zippers. Nature 447:453–457

Schutz AK, Soragni A, Hornemann S, Aguzzi A, Ernst M, Bockmann A, Meier BH (2011) The amyloid-Congo Red interface at atomic resolution. Angew Chem Int Ed Engl 50:5956–5960

Shewmaker F, Wickner RB, Tycko R (2006) Amyloid of the prion domain of Sup35p has an in-register parallel β-sheet structure. Proc Natl Acad Sci USA 103:19754–19759

Sievers SA, Karanicolas J, Chang HW, Zhao A, Jiang L, Zirafi O, Stevens JT, Munch J, Baker D, Eisenberg D (2011) Structure-based design of non-natural amino acid inhibitors of amyloid fibril formation. Nature 475:96–100

Sunde M, Serpell LC, Bartlam M, Fraser PE, Pepys MB, Blake CCF (1997) Common core structure of amyloid fibrils by synchrotron X-ray diffraction. J Mol Biol 273:729–739

Telling GC, Parchi P, DeArmond SJ, Cortelli P, Montagna P, Gabizon R, Mastrianni J, Lugaresi E, Gambetti P, Prusiner SB (1996) Evidence for the conformation of the pathologic isoform of the prion protein enciphering and propagating prion diversity. Science 274:2079–2082

Tjernberg LO, Callaway DJE, Tjernberg A, Hahne S, Lilliehook C, Terenius L, Thyberg J, Nordstedt C (1999) A molecular model of Alzheimer amyloid β-peptide fibril formation. J Biol Chem 274:12619–12625

Torok M, Milton S, Kayed R, Wu P, McIntire T, Glabe CG, Langen R (2002) Structural and dynamic features of Alzheimer's Aβ peptide in amyloid fibrils studied by site-directed spin labeling. J Biol Chem 277:40810–40815

Toyama BH, Kelly MJS, Gross JD, Weissman JS (2007) The structural basis of yeast prion strain variants. Nature 449:233–237

Tycko R (2006) Molecular structure of amyloid fibrils: insights from solid state NMR. Q Rev Biophys 39:1–55

Tycko R (2011) Solid state NMR studies of amyloid fibril structure. Annu Rev Phys Chem 62:279–299

Tycko R, Hu K-N (2010) A Monte Carlo/simulated annealing algorithm for sequential resonance assignment in solid state NMR of uniformly labeled proteins with magic-angle spinning. J Magn Reson 205:304–314

Tycko R, Sciarretta KL, Orgel J, Meredith SC (2009) Evidence for novel β-sheet structures in Iowa mutant β-amyloid fibrils. Biochemistry 48:6072–6084

Tycko R, Savtchenko R, Ostapchenko VG, Makarava N, Baskakov IV (2010) The α-helical C-terminal domain of full-length recombinant PrP converts to an in-register parallel β-sheet structure in PrP fibrils: evidence from solid state nuclear magnetic resonance. Biochemistry 49:9488–9497

Van Melckebeke H, Wasmer C, Lange A, Eiso AB, Loquet A, Bockmann A, Meier BH (2010) Atomic-resolution three-dimensional structure of HET-s218-289 amyloid fibrils by solid state NMR spectroscopy. J Am Chem Soc 132:13765–13775

Wasmer C, Soragni A, Sabate R, Lange A, Riek R, Meier BH (2008) Infectious and noninfectious amyloids of the HET-s218-289 prion have different NMR spectra. Angew Chem Int Ed Engl 47:5839–5841

Wickner RB, Dyda F, Tycko R (2008) Amyloid of Rnq1p, the basis of the [PIN+] prion, has a parallel in-register β-sheet structure. Proc Natl Acad Sci USA 105:2403–2408

Wong DF, Rosenberg PB, Zhou Y, Kumar A, Raymont V, Ravert HT, Dannals RF, Nandi A, Brasic JR, Ye WG, Hilton J, Lyketsos C, Kung HF, Joshi AD, Skovronsky DM, Pontecorvo MJ (2010) In vivo imaging of amyloid deposition in Alzheimer disease using the radioligand 18F-AV-45 (Flobetapir F 18). J Nucl Med 51:913–920

Structure-Activity Relationship of Amyloids

Jason Greenwald and Roland Riek

Abstract Amyloids are highly ordered, cross-β-sheet protein aggregates. The unique cross-β-sheet entity is composed of an indefinitely repeating inter-molecular β-sheet motif. It can grow by recruitment of the corresponding amyloid protein, while its repetitiveness can translate what would be a non-specific activity as monomer into a potent one through cooperativity. Because of these properties, the activities of amyloids are manifold and include peptide storage, template assistance, loss of function, gain of function, generation of toxicity, membrane binding, infectivity, etc. Thus, amyloids are associated both with diseases, including Alzheimer's, Creutzfeldt-Jakob and Parkinson's disease, and biological functions such as hormone storage in secretory granules and skin pigmentation in mammals. This review summarizes the recent high-resolution structural studies of amyloid fibrils in light of their biological activities, with special focus on the functional HET-s prion system and hormone storage in secretory granules.

Protein aggregation into amyloids is associated both with dozens of diseases, including Alzheimer's, Parkinson's and prion diseases (Chiti and Dobson 2006), and with biological function such as skin pigmentation or hormone storage (Fowler et al. 2007; Maji et al. 2009). We adopt the modern biophysical definition of an amyloid fibril being an unbranched protein fiber whose repeating substructure consists of β-strands that run perpendicular to the fiber axis, forming an indefinite intermolecular cross-β-sheet (Fig. 1; Greenwald and Riek 2010; Sipe and Cohen 2000; Astbury et al. 1935; Sunde et al. 1997). Among protein folds, this structural entity is unique. In particular, it can grow by recruitment of the corresponding amyloid protein, and its repetitiveness can translate what would be a non-specific activity as monomer into a potent one through cooperativity, resulting in many activities (both good and bad for

J. Greenwald • R. Riek (✉)
ETH Zurich, Physical Chemistry, ETH Honggerberg, HCI F 225, Wolfgang-Pauli-Str., 108093 Zurich, Switzerland
e-mail: roland.riek@phys.chem.cthz.ch

M. Jucker and Y. Christen (eds.), *Proteopathic Seeds and Neurodegenerative Diseases*, Research and Perspectives in Alzheimer's Disease, DOI 10.1007/978-3-642-35491-5_3, © Springer-Verlag Berlin Heidelberg 2013

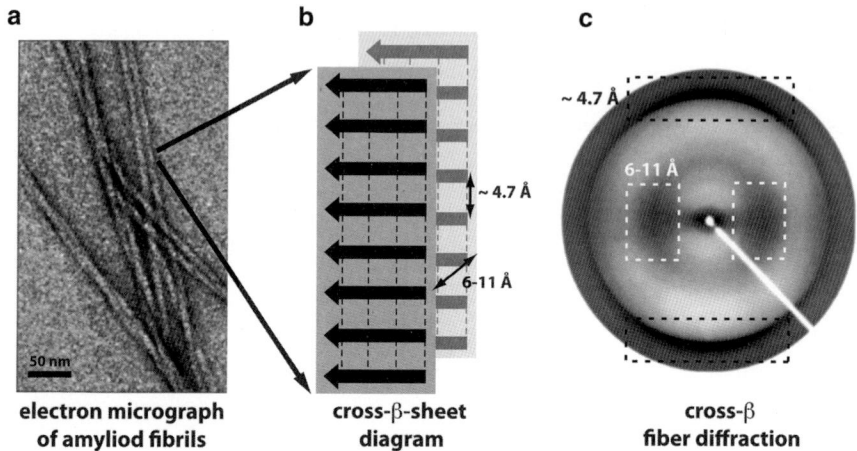

a   **b**   **c**

electron micrograph   cross-β-sheet   cross-β
of amyliod fibrils   diagram   fiber diffraction

Fig. 1 (**a**) Amyloid fibrils are composed of long filaments that are visible in negatively stained transmission electron micrographs. (**b**) The schematic diagram of the cross-β-sheets in a fibril, with the backbone hydrogen bonds represented by *dashed lines*, indicates the repetitive spacings that give rise to (**c**), the typical fiber diffraction pattern with a meridional reflection at ~4.7 Å (*black dashed box*) and an equatorial reflection at around 6–11 Å (*white dashed box*)

the cell). It is the aim of this review to highlight the structure-activity relationship of two functional amyloid systems: the HET-s prion and pituitary hormone storage in secretory granules.

Amyloid Crystal Structures of Short Amyloidogenic Peptides: The Cross-β Spine at Atomic Resolution

By using short fibril-forming peptide segments of amyloid proteins, researchers in the Eisenberg lab were able to grow 3D microcrystals that likely represent the structure in the fibril form (i.e., in the 1D crystal-like form; Ivanova et al. 2009; Nelson et al. 2005; Sawaya et al. 2007; Wiltzius et al. 2008, 2009). The structures represent a survey of the types of interactions that can support a cross-β core, including the β-strand and β-sheet symmetries that can exist in a cross-β fibril. For example, the collection of structures reveals two classes of β-sheet stacking interfaces, termed the "dry" and "wet" interfaces (Fig. 2). The "dry" interface is devoid of water molecules, consisting of complementary side chain interdigitation that results in a high peptide packing density. Such a side chain interdigitation is termed a "steric zipper" motif. The "wet" interface is composed of hydrogen bonds between side chains, some via water molecules, in a manner similar to intermolecular contacts in protein crystals. This finding, along with the fact that some peptides have been found in multiple crystal forms with the same "dry" but different "wet"

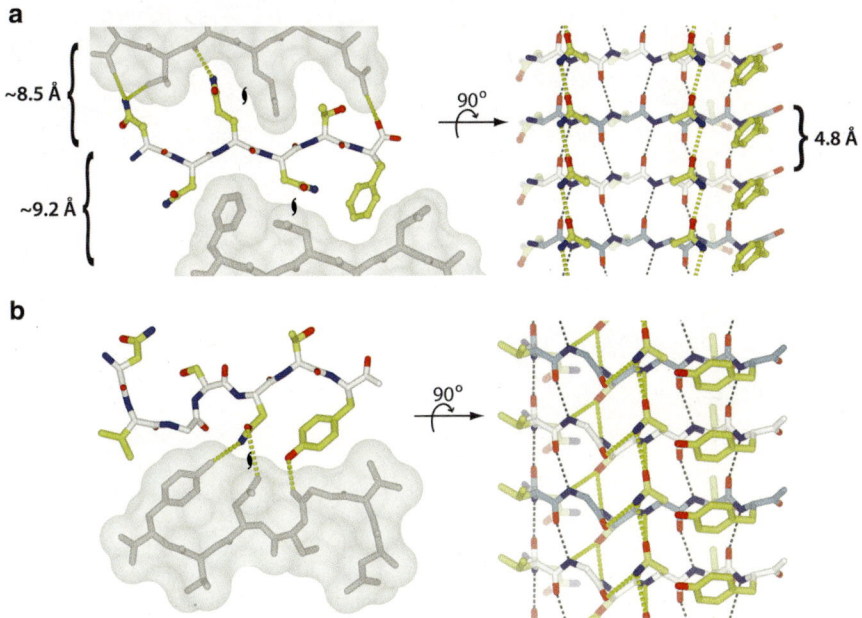

Fig. 2 Steric zippers, hydrogen bonding and van der Waals packing in atomic resolution amyloid crystal structures. The interactions found in cross-β fibrils are exemplified by the structures of two peptides: (**a**) NNQNTF (PDB code: 3FVA) and (**b**) NVGSNTY (PDB code: 3FTL). The *left panels* are the views looking down the axes of the microcrystal "fibrils," with the neighboring dry interfaces depicted as a solvent-accessible surface (*gray*) to show the tight interdigitation of the side chains. The inter-sheet hydrogen bonds are shown as *yellow dashed lines* (only a unique set is shown, not symmetry equivalent bonds). The coloring scheme is *white* for main-chain carbon, *yellow* for side chain carbon, *blue* for nitrogen and *red* for oxygen. The positions of the twofold screw axes that relate the individual molecules in the fibril are indicated with *black hurricane symbols*. The *right panels* are the view from the side of the "fibrils" showing only the β-sheet corresponding to the colored strand from the *left panel*. The intra-sheet hydrogen bonds are shown (*black* for main-chain only bonds and *yellow* for any side chain interactions), and the coloring of the main-chain carbons of individual strands alternates *blue* and *white*. The front face of the sheet in (**a**) has two Asn ladders and a Phe ladder. The peptide in (**b**) is actually composed of two short β-strands with a kink between them. The kink allows the Asn in the ladder to make a hydrogen bond to the main-chain oxygen of residue three (Gly) within the same molecule. In (**a**), the spacing between the repetitive structures is indicated (the *left panel* has only the average spacing between main-chain atoms because the inter-sheet distances are not related by crystallographic translations along unit cells)

interfaces, suggests that the stable structural unit of the microcrystals, and hence of the fibrils, is a pair of β-sheets. Although the amyloids of full-length proteins should have more complex structures, the types of interactions that stabilize them are likely to be the same as those observed in the short peptides, as exemplified in the following.

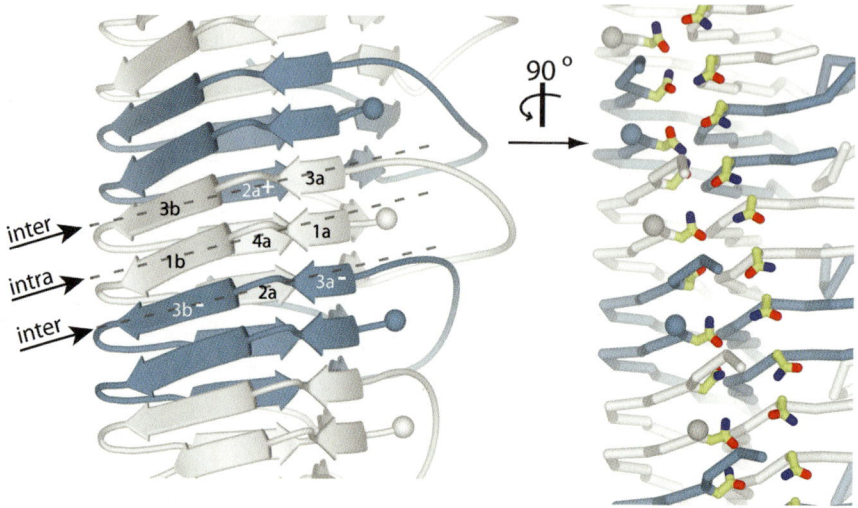

Fig. 3 The β-solenoid motif in HET-s. The *left view* is a ribbon representation of the HET-s PFD fibril with the peptide chains alternating *blue* and *white* and a ball on the N-termini (residue 225). The 4 β-strands are numbered and with "a" and "b" designations for the co-directional strands that are broken by a short kink and "−" and "+" designations for the strands from the molecule below and above the *central white one*. In both views, it is clear the side chain interactions between the two sheets occur between strands 1 and 4 of the same molecule and between strands 2 and 3 of neighboring subunits. The *right view* is 90° rotated from the *left* and is showing only the Cβ trace of the cross-β core and the 4 Asn side chains in the interior of the core that form a double Asn ladder. The absence of the visible hydrogen-bond network in the Asn ladder reflects the limitations of the technique and refinement method rather than the actual conformation of the Asn side chains

The Bigger Picture: Amyloid Fibrils of HET-s(218–289)

HET-s is a functional prion from the filamentous fungi *Podospora anserina* that is involved in a self-nonself discrimination process (Coustou et al. 1997). Like most prion proteins, HET-s is a multi-domain protein. It has a globular N-terminal domain and a flexible and disordered C-terminal prion-forming domain (PFD; Balguerie et al. 2003). The PFD (comprising residues 218–289) can undergo a large structural rearrangement to form a stable amyloid that is itself the infectious entity, able to induce other HET-s molecules to form the amyloid structure. The 3D structure of HET-s(218–289) fibrils was determined by solid state NMR (Wasmer et al. 2008). The fibril is a left-handed β-solenoid with each protein molecule forming two helical turns. Because there are two helical windings per molecule, the β-strands alternate between inter- and intra-molecular hydrogen bonding along the fiber axis. In addition, there is an offset in the winding of the solenoid so that the side chain packing between the sheets is both inter- and intra-molecular (Fig. 3), increasing the size and complexity of the intermolecular interface [note: a similar intermolecular side chain arrangement has been proposed for the A β(1–42) fibrils associated with Alzheimer's disease (Luhrs et al. 2005)]. In contrast to all of the small peptide structures in which none of the inter-strand contacts are intra-molecular, the HET-s fibril β-strands pack with ~50 % intra-molecular contacts (inter-strand and inter-sheet). Furthermore, the pseudo-repeat in the PFD amino acid sequence generates a pattern of alternating charges along the fibril axis that supports the correct in-register alignment of the β-sheets (Fig. 3). Thus, it appears that the HET-s monomer contains structural elements that facilitate amyloid fibril nucleation (pseudo-repeats), growth (winding offset; Ritter et al. 2005) and "in register" intermolecular alignment (alternate charged side chains). Within this context, it is worth mentioning that, as a functional amyloid, HET-s has evolved to fold into a cross-β motif and therefore may be more complex than the disease-related amyloids for which the cross-β structure is an unfortunate energy minimum on the folding landscape. The β-solenoid is in fact a large structural family of soluble proteins (Kajava and Steven 2006) and the complexity of the HET-s PFD amyloid structure is on a par with other protein folds. It is ironic though, that with its mere two helical turns per molecule, the HET-s PFD is the shortest β-solenoid in sequence but a structure that can form fibrils longer than 100 μm.

Despite its complexity, close inspection reveals that the HET-s fibril consists of the same structural features that are found in the microcrystals: a hydrophobic core, steric zipper-like interactions for the inter-sheet packing, and Asn ladders and pi-pi stacking running along the length of the fiber (Fig. 3; note, however, that solid-state NMR and distance restraint-based refinements in general cannot generate models with the accuracy required to assign the conformation of every side chain).

The Amyloid Compared to Other Protein Aggregates

Although the amyloid-like conformation is a common feature of many protein aggregates, there are other mechanisms by which proteins can assemble into ordered aggregates without invoking the cross-β motif. These mechanisms include domain swapping, end-to-end stacking (Bennett et al. 2006; Eisenberg et al. 2006; Nelson and Eisenberg 2006a, b), and silk-type β-sheet polymerization. The functional aggregation of proteins in an end-to-end or lateral manner is most well studied in the cytoskeletal complexes of the actins and tubulins. Their correct function requires that there be a tight control over the assembly and disassembly of their ordered aggregates (i.e., actin filaments and microtubules). In the case of actin, the filaments retain the ability to make branch points that are critical for the rigidity of the cytoskeleton by incorporating specialized branch-inducing complexes into the fila-ment. Because it does not involve a gross rearrangement of the protein fold, the end-to-end type of aggregation is reversible and does not limit much the speed at which the assembly and disassembly occur. The regulation of the assembly process is similar for both actin and the tubulins in which nucleating factors stimulate the initiation of aggregation while nucleoside triphosphate (NTP) binding and hydroly-sis regulates the assembly/disassembly. Another example of end-to-end aggregation is the assembly of hemoglobin S (HbS), the glutamate to valine point mutant of hemoglobin A, into end-to-end and lateral fibers. The fibers of HbS distort the red blood cells into the abnormal rigid shape that gives the name to the resulting disease: sickle cell anemia. The polymerization of HbS is affected by several factors, including oxygen saturation, hemoglobin concentration and hemoglobin composition. This latter effect spares HbS/HbA individuals from severe anemia and, since there is no biological control over the assembly/disassembly, HbS/HbS individuals are afflicted to various degrees.

Contrasting with the end-to-end/lateral aggregates are those of the arthropod silks, whose ordered assembly is rapid and obtained on demand from specialized glands. The enormous variety of silks represents many uncharacterized mechanisms for polypeptide assembly, yielding structures that range from primarily α-helical content to parallel β-sheets and even to cross-β amyloid-like structures. However, the more well-studied spider fibroins present an interesting mechanism that involves dynamic control like the end-to-end aggregates, a large and irreversible structural rearrangement like the amyloid aggregates and a specialized production machinery like the keratin/collagen aggregates. The fibroins have multiple repeating domains of several types, with highly conserved terminal non-repeating (NR) globular domains. Protein constructs that contain just two types of repeats (Gly/Ala rich and Gly/Pro/Gln rich) can form a β-rich aggregate, whereas the addition of the C-terminal NR domain to the construct produces a protein that in vitro displays many of the aggregation properties of silk. The different types of repeating domains give rise to the many properties of silk, most importantly its tensile strength and flexibility. Upon silk fiber assembly, the Gly/Ala rich regions (thought to be helical in solution) form a β-sheet-rich structure in which, unlike amyloids, the strands run parallel to the fiber axis, whereas the Gly/Pro/Gln segments adopt an uncharacterized structure that gives the elasticity to the silk. A recent study has determined the structure of the C-terminal non-repeating domain from the dragline silk of *Araneus diadematus* and showed that it is essential for controlled aggregation and ordered fiber assembly (Hagn et al. 2010). The current data support the hypothesis that it is the terminal NR domains that allow for the storage of the aggregation prone repeats at a very high concentration in the spinning gland (up to 50 % w/v). The conversion from the soluble proteins (thought to be in either a liquid crystal or micellar suspension) into the ordered insoluble fibers is stimulated by a change in the salt type (from Cl^- to PO_4^{3-}), a decrease in pH, dehydration and by the shear forces as the proteins pass through the narrow spinning duct. Thus it appears that both a prior ordering and a mechanical force are required to achieve the ordered assembly of fibroins into a dragline silk fiber.

Yet another type of protein aggregation occurs in the assembly of the structural proteins like collagen and the keratins. These structural proteins comprise a variety of folds that we will not discuss here except to say that what separates them from other ordered aggregates is the fact that they do not spontaneously form ordered aggregates; rather, they require a large degree of specialized cellular machinery and processing to achieve their final state. Finally, a group of aggregates that are classified solely on their macroscopic appearance are the amorphous aggregates, a type that is often obtained by heat precipitation or concentration precipitation of proteins. While the classical macroscopic amorphous aggregate is on the mesoscopic scale different from the fibrous aggregate, a recent study from our lab indicates that amorphous aggregates may also contain a defined cross-β-structure, but without the long range (>μm) order found in fibrils (Wang et al. 2010b). This finding suggests that the manner in which a polypeptide backbone achieves a kinetically accessible local energy minimum might require that there is some local order or repeated structure throughout the aggregate. Thus, there appears to be a caveat about using the terms "ordered" versus "amorphous" protein aggregates, and it is perhaps more correct to describe aggregates based on an order/disorder continuum or an extent of long-range ordering (i.e., coherence length).

In summary, a comparison of the various types of aggregates described above reveals two significant features that set the amyloid apart from other ordered aggregates. First, the cross-β spine of the amyloid can be formed from peptides as short as four residues. Second, and partly due to the first, the repeat distance is short; in the most basic amyloid, it is the inter-strand distance of 4.7 Å. The close spacing of identical side chains can generate specificity where none would exist without a repeating structure. For example, a single amphipathic β-strand would have no affinity for a membrane, but a fibril made up of this peptide could bind tightly to lipids. The same is true of a basic peptide interacting with DNA (see below). In addition to supporting cooperative binding, the amyloid structure allows very short peptides to assemble into complexes with regular secondary structure, thus mimicking some features of larger folded proteins. Hence, amyloids have unique structural properties that can result in unique activities, as we shall see in the following section.

Structure-Activity Relationship

The two-state (soluble/insoluble) nature of amyloids and the structural repetitiveness of the aggregates add a level of complexity that most soluble proteins cannot access. Hence, amyloids can possess a variety of biophysical and biological properties, supporting a diversity of functions that rivals that of soluble proteins. The growing list of known amyloid related activities shows that the formation of amyloids may result both in a loss of function of the polypeptide, as described for the yeast prions (Osherovich and Weissman 2002; True and Lindquist 2000), or in a gain of function, as in the case of the HET-s prion (Maddelein et al. 2002). Pmel17 amyloids serve as a template for ligand binding (Fowler et al. 2006), whereas other amyloids induce a specific toxic response, as shown for the HET-s prion system (Maddelein et al. 2002). The structural repetitiveness of the amyloid fiber provides an ideal template for making it into a transmissible or infectious fold. In fact, most prions (infectious proteins) have been found in their infectious form to be amyloids whereas the amyloid structure itself appears to be responsible for prion strain diversity (Toyama et al. 2007). Although, the in vitro formation of most amyloids can be nucleated or "seeded" by the amyloid itself from a soluble pool of the amyloidogenic peptide or protein, not all amyloids have been shown to be transmissible and thus not all amyloids can be considered prions (Riek 2006). Thus the simplicity of a cross-β-structure with its one-dimensional crystal-like nature can encompass a wide range of activities. As we will highlight in the following specific examples, their activity can be dependent on the repetitive nature of the amyloid but also on its two-state nature and the compactness of the fold.

The Structure-Activity Relationship in the HET-s Amyloid Prion

The structure of the prion-forming domain (PFD) from HET-s revealed that the infectious state of this functional amyloid is a β-solenoid and thus a member of a large structural class of soluble proteins (Maddelein et al. 2002; Ritter et al. 2005; Wasmer et al. 2008). The complexity of its structure reflects its evolutionarily optimized role as an amyloid. The β-solenoid structure is not a chance misfolding of the PFD but rather a very stable conformation that can spontaneously occur at moderate protein concentration. The presence of a pseudo-repeat in the sequence allows a single PFD to adopt two turns of the β-solenoid, forming the nucleus of the amyloid fold. As well, it gives rise to alternating charges along the fibril axis that support the correct in-register alignment of the β-sheets (Fig. 3). The pairing of these repeats guarantees that there is a unique low-energy fold without detectable polymorphism in physiological HET-s fibrils (Wasmer et al. 2008). With its hydrophobic core and hydrophilic solvent-exposed side chains, the HET-s PFD fibril has many of the hallmarks of soluble proteins. In fact, unlike many amyloidogenic proteins, there is no toxicity associated with its over-expression in bacterial or fungal hosts or even in rats (Winner et al. 2011).

It had been previously shown that the introduction of the β-sheet-breaking residue proline within the predicted β-strands resulted in a loss of infectivity in the fungus, whereas proline mutations in the loop positions had no effect (Ritter et al. 2005). Therefore, the high-resolution structure was able to confirm that the amyloid core is the infectious entity. However, of its many functions, only the infectivity of the HET-s prion is completely contained within its PFD. The heterokaryon incompatibility function requires a full-length protein that includes the N-terminal globular domain (termed the HeLo domain based on its conservation with other fungal proteins) and the PFD (Balguerie et al. 2004). Furthermore, the HeLo domain has a prion regulatory function that also requires a full-length protein with the proper spacing between the domains (Balguerie et al. 2003; Greenwald et al. 2010). In the latter report, we conclude that the formation of the amyloid structure in the PFD causes a structural rearrangement of the HeLo domain that leads to the other activities of the HET-s prion. In this manner, the combination of a globular fold and an amyloid could create an immensurable functional landscape and one that has certainly been explored by nature and that is awaiting discovery in the laboratory. The evolutionarily optimized fibril structure of HET-s has many features that are likely to be found in other highly evolved functional amyloids (Barnhart and Chapman 2006; Fowler et al. 2006, 2007).

Peptide Hormone Amyloids in the Formation of Secretory Granules

Our recent finding that the amyloid structure is involved in the formation of secretory granules in mammalian cells (Maji et al. 2009) hints at the biological ubiquity of functional amyloids. Eukaryotic cells have specialized pathways for both the constitutive and regulated transport of secretory proteins/peptides to the extracellular space (Kelly 1985, 1987). During constitutive secretion, a newly synthesized peptide hormone is transported to the cell surface whereupon it is immediately released and the rate of peptide secretion is limited by the rate of peptide synthesis (Arvan and Castle 1998; Lacy 1975; Palade 1975; Tooze 1998). In contrast, many secretory cells have an additional secretion pathway through which they can store peptide hormones for extended periods of time. By packaging the secretory peptides in highly concentrated membrane-enclosed secretory granules (Arvan and Castle 1998; Dannies 2001; Kelly 1985), the cells can store a large quantity of the hormone until a signal triggers its release, at which point they can secrete hormones much faster than their synthesis rates would permit.

During the maturation process, the prohormones are proteolytically processed into their active hormone entity followed by hormone aggregation into hormone-specific granules that are membrane-enclosed. The granules are generally composed of a single peptide species or, when co-aggregated, the distinct peptides exist exclusively in a specific integer ratio. Until recently, there was no explanation for how peptide hormones self-segregated, but the amyloid nature of peptide hormones in secretory granules (Maji et al. 2009) explains much of what were mysteries in the processes of granule formation, selection, storage, and hormone release. The nature of amyloids provides a series of controls over secretory granules that are semi-autonomous, requiring minimal peripheral cell machinery (Fig. 4).

1. The timing and location of amyloid aggregation in the Golgi can be controlled by the processing of the prohormone. Since the amyloid aggregation of a small peptide is sensitive to the sequence and chemical modifications of the peptide, there are multiple ways in which processing could initiate aggregation. In addition to controlling aggregation through modification of the hormone, there is the possibility of controlling aggregation through pH or the presence of helper molecules such as glycosaminoglycans (GAGs) that can stabilize the amyloid conformation (Maji et al. 2009).

2. The formation of amyloid fibrils is highly sequence-specific so that, once initiated, the amyloid aggregation of the hormone is self-selective, yielding granule cores composed of specific hormones only. Specific coaggregation of some hormones is possible, as has been demonstrated for ACTH and β-endorphin, two hormones that are processed from a single prohormone. Also, since some amyloid proteins are able to cross-seed (Giasson et al. 2003; Han et al. 1995), it should be expected that some hormones will form mixed fibrils, although still in a specific manner.

3. After initiation of aggregation and sorting of the peptides into specific secretory granule cores, the amyloid structure provides the most dense packing possible

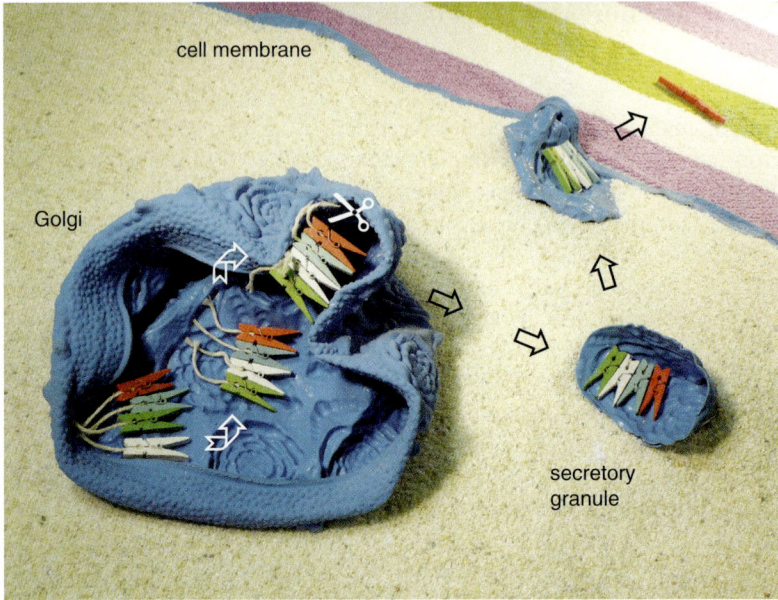

Fig. 4 The amyloid activities in the secretory granule biogenesis. The amyloid state of peptide hormones in secretory granules (shown by an alignment of several clothes pins) may explain the processes of granule formation, selection, storage, and release of hormones in the granules. It is suggested that, in the Golgi (shown as a *blue bathing cup*), the amyloid aggregation of the pro-hormone (shown by individually colored clothespins with a tail) is initiated spontaneously above a critical pro-hormone concentration or/and in presence of helper molecules such as glycosami-noglycans in parallel to a possible pro-hormone processing (indicated by a *scissor*) that also may initiate the aggregation. Since the formation of amyloid fibrils is highly amino acid sequence-specific, the initiated amyloid aggregation of the (pro)-hormone is selective, excluding non-aggregation-prone constitutively secreted proteins and yielding granule cores composed of single hormones or multiple distinct hormone co-aggregates only. Since the amyloid entity is usually able to interact with membranes (shown in *blue*), the hormone amyloid is wrapped into membranes during the aggregation process, followed by the formation of granules. The mature secretory granules can exist for long durations since amyloids are very stable. Upon stimulation, secretory granules are secreted and release monomeric, functional hormone (indicated by an extended *red stick*) in a controlled manner. The *scissor* indicates the convertases. The cytoplasm is shown as *sand*, the extracellular space as a bath-towel. The *red colored stick* represents the released soluble hormone

(Nelson et al. 2005), both excluding non-aggregation-prone constitutively secreted proteins and providing an extremely stable storage system.

4. As part of the granule maturation process, the hormone amyloids are surrounded by membrane as they separate from the Golgi. There are many examples to suggest that membrane binding is an inherent property of amyloid aggregates (Sparr et al. 2004), so formation of a membrane around the hormone aggregate may be spontaneous.

5. Finally, upon receiving the appropriate signal, the granules are secreted and the amyloid structure of the hormone aggregates enables a controlled release of

monomeric, functional hormone (Maji et al. 2008). Each peptide hormone will have its own dissociation rate and this rate can again be controlled by factors such as pH, ion concentration, and extracellular chaperones.

Distinct Aggregates, Distinct Activities

Our recent structural studies on HypF-N aggregates show that even the rapid, forced aggregation of a protein can lead to several distinct amyloid-like states whose conformations depend on the conditions in which they were formed (Wang et al. 2010b). Furthermore, the activities of the aggregates in several biochemical and biological assays varied with the conformation of the aggregates. HypF-N is the N-terminal domain of the prokaryotic hydrogenase maturation factor, which has been shown to form toxic oligomers on way to forming amyloid fibrils. In addition to these amyloid fibrils, we studied the bacterial inclusion bodies, heat-precipitated aggregates, concentration-induced aggregates, and trichloroacetic acid-precipitated aggregates of HypF-N. Despite the wide range of conditions used to induce protein aggregation, all five types of HypF-N aggregates contain the cross-β-sheet motif. However, each of the aggregates is structurally distinct, having different segments of the protein involved in protected secondary structures. We compared the affinity of each aggregate to ATP, Thioflavin T, oligonucleotides, and micelles, as well as the ability of each to interfere with cell viability. Although the highest activities were found in the fibrils, the other aggregates had significant activities that were not inter-correlated (e.g., strong ATP binding was not correlated with cell viability). We believe that the key to their varying properties is that each aggregate is built around a cross-β-sheet motif composed of different amino acid segments. The repetitive nature of the cross-β-sheet motif supports cooperative interactions that can generate unique and potent activities. This one example of the landscape of aggregate structure-activity relationships highlights the existence of a multi-factorial structure-activity function for amyloids and begins to shed light on the complexities of these polymorphic aggregates.

Conclusion

In 1935, the pioneering biophysicist Astbury published an X-ray diffraction pattern from poached, stretched egg white that displayed the reflections typical for the cross β-sheet entity. The authors concluded that, when proteins/peptides aggregate, they go into their energetically most favorable conformational state, this being the cross-β-sheet motif. In other words, when a protein aggregates it generally forms an amyloid-like entity comprising a specific structure. Protein aggregation can thus be viewed as a primitive folding process that results in a defined set of aggregated conformations, which makes the amyloid entity a prime candidate both for the origin of life and for the first folds of the known protein universe. Furthermore, it is consistent with the fact that there are many functional amlyoids that are involved in diverse mechanisms such as protection, storage, infectivity and perhaps long-term memory. Although 75 years have passed since the amyloid was proposed as a common structural motif, the structural and functional characterization of the amyloid world is just in the starting phase of discovery, with much yet to be revealed.

Acknowledgment This review is adopted from a review written by Greenwald and Riek (2010) and from a review written by Seuring et al. (2012). The artist of Fig. 4 is Christina Comiotto.

References

Arvan P, Castle D (1998) Sorting and storage during secretory granule biogenesis: looking backward and looking forward. Biochem J 332:593–610

Astbury WT, Dickinson S, Bailey K (1935) The X-ray interpretation of denaturation and the structure of the seed globulins. Biochem J 29:2351–2360

Balguerie A, Dos Reis S, Ritter C, Chaignepain S, Coulary-Salin B, Forge V, Bathany K, Lascu I, Schmitter JM, Riek R, Saupe SJ (2003) Domain organization and structure-function relationship of the HET-s prion protein of Podospora anserina. EMBO J 22:2071–2081

Balguerie A, Dos Reis S, Coulary-Salin B, Chaignepain S, Sabourin M, Schmitter JM, Saupe SJ (2004) The sequences appended to the amyloid core region of the HET-s prion protein determine higher-order aggregate organization in vivo. J Cell Sci 117:2599–2610

Barnhart MM, Chapman MR (2006) Curli biogenesis and function. Annu Rev Microbiol 60:131–147

Bennett MJ, Sawaya MR, Eisenberg D (2006) Deposition diseases and 3D domain swapping. Structure 14:811–824

Chiti F, Dobson CM (2006) Protein misfolding, functional amyloid, and human disease. Annu Rev Biochem 75:333–366

Coustou V, Deleu C, Saupe S, Begueret J (1997) The protein product of the HET-s heterokaryon incompatibility gene of the fungus Podospora anserina behaves as a prion analog. Proc Natl Acad Sci USA 94:9773–9778

Dannies PS (2001) Concentrating hormones into secretory granules: layers of control. Mol Cell Endocrinol 177:87–93

Eisenberg D, Nelson R, Sawaya MR, Balbirnie M, Sambashivan S, Ivanova MI, Madsen AO, Riekel C (2006) The structural biology of protein aggregation diseases: fundamental questions and some answers. Acc Chem Res 39:568–575

Fowler DM, Koulov AV, Alory-Jost C, Marks MS, Balch WE, Kelly JW (2006) Functional amyloid formation within mammalian tissue. PLoS Biol 4:e6

Fowler DM, Koulov AV, Balch WE, Kelly JW (2007) Functional amyloid – from bacteria to humans. Trends Biochem Sci 32:217–224

Giasson BI, Forman MS, Higuchi M, Golbe LI, Graves CL, Kotzbauer PT, Trojanowski JQ, Lee VMY (2003) Initiation and synergistic fibrillization of tau and a-synuclein. Science 300:636–640

Greenwald J, Riek R (2010) Biology of amyloid: structure, function, and regulation. Structure 18:1244–1260

Greenwald J, Buhtz C, Ritter C, Kwiatkowski W, Choe S, Maddelein ML, Ness F, Cescau S, Soragni A, Leitz D, Saupe S, Riek R (2010) The mechanism of prion inhibition by HET-S. Mol Cell 38:889–899

Hagn F, Eisoldt L, Hardy JG, Vendrely C, Coles M, Scheibel T, Kessler H (2010) A conserved spider silk domain acts as a molecular switch that controls fibre assembly. Nature 465:239–242

Han HY, Weinreb PH, Lansbury PT (1995) The core Alzheimer's peptide NAC forms amyloid fibrils which seed and are seeded by b-Amyloid – is NAC a common trigger or target in neurodegenerative disease? Chem Biol 2:163–169

Ivanova MI, Sievers SA, Sawaya MR, Wall JS, Eisenberg D (2009) Molecular basis for insulin fibril assembly. Proc Natl Acad Sci USA 106:18990–18995

Kajava AV, Steven AC (2006) b-Rolls, b-helices, and other b-solenoid proteins. Adv Protein Chem 73:55–96

Kelly RB (1985) Pathways of protein secretion in eukaryotes. Science 230:25–32

Kelly RB (1987) Protein transport. From organelle to organelle. Nature 326:14–15

Lacy PE (1975) Endocrine secretory mechanisms. A review. Am J Pathol 79:170–188

Luhrs T, Ritter C, Adrian M, Riek-Loher D, Bohrmann B, Dobeli H, Schubert D, Riek R (2005) 3D structure of Alzheimer's amyloid-b(1-42) fibrils. Proc Natl Acad Sci USA 102:17342–17347

Maddelein ML, Dos Reis S, Duvezin-Caubet S, Coulary-Salin B, Saupe SJ (2002) Amyloid aggregates of the HET-s prion protein are infectious. Proc Natl Acad Sci USA 99:7402–7407

Maji SK, Schubert D, Rivier C, Lee S, Rivier JE, Riek R (2008) Amyloid as a depot for the formulation of long-acting drugs. PLoS Biol 6:e17

Maji SK, Perrin MH, Sawaya MR, Jessberger S, Vadodaria K, Rissman RA, Singru PS, Nilsson KP, Simon R, Schubert D, Eisenberg D, Rivier J, Sawchenko P, Vale W, Riek R (2009) Functional amyloids as natural storage of peptide hormones in pituitary secretory granules. Science 325:328–332

Nelson R, Eisenberg D (2006a) Recent atomic models of amyloid fibril structure. Curr Opin Struct Biol 16:260–265

Nelson R, Eisenberg D (2006b) Structural models of amyloid-like fibrils. Adv Protein Chem 73:235–282

Nelson R, Sawaya MR, Balbirnie M, Madsen AO, Riekel C, Grothe R, Eisenberg D (2005) Structure of the cross-b spine of amyloid-like fibrils. Nature 435:773–778

Osherovich LZ, Weissman JS (2002) The utility of prions. Dev Cell 2:143–151

Palade G (1975) Intracellular aspects of the process of protein synthesis. Science 189:347–358

Riek R (2006) Cell biology: infectious Alzheimer's disease? Nature 444:429–431

Ritter C, Maddelein ML, Siemer AB, Luhrs T, Ernst M, Meier BH, Saupe SJ, Riek R (2005) Correlation of structural elements and infectivity of the HET-s prion. Nature 435:844–848

Sawaya MR, Sambashivan S, Nelson R, Ivanova MI, Sievers SA, Apostol MI, Thompson MJ, Balbirnie M, Wiltzius JJ, McFarlane HT, Madsen AO, Eisenberg D (2007) Atomic structures of amyloid cross-b spines reveal varied steric zippers. Nature 447:453–457

Seuring S, Nespovitaya N, Rusihauser J, Spiess M, Riek R (2012) Hormone amyloids in sickness and in health. In: Otzen D (ed) Amyloid fibrils and prefibrillar aggregates molecular and biological properties. Wiley-VCH, Weinheim

Sipe JD, Cohen AS (2000) Review: history of the amyloid fibril. J Struct Biol 130:88–98

Sparr E, Engel MF, Sakharov DV, Sprong M, Jacobs J, de Kruijff B, Hoppener JW, Killian JA (2004) Islet amyloid polypeptide-induced membrane leakage involves uptake of lipids by forming amyloid fibers. FEBS Lett 577:117–120

Sunde M, Serpell LC, Bartlam M, Fraser PE, Pepys MB, Blake CC (1997) Common core structure of amyloid fibrils by synchrotron X-ray diffraction. J Mol Biol 273:729–739

Tooze SA (1998) Biogenesis of secretory granules in the trans-Golgi network of neuroendocrine and endocrine cells. Biochim Biophys Acta 1404:231–244

Toyama BH, Kelly MJ, Gross JD, Weissman JS (2007) The structural basis of yeast prion strain variants. Nature 449:233–237

True HL, Lindquist SL (2000) A yeast prion provides a mechanism for genetic variation and phenotypic diversity. Nature 407:477–483

Wang L, Schubert D, Sawaya MR, Eisenberg D, Riek R (2010b) Multidimensional structure-activity relationship of a protein in its aggregated states. Angew Chem Int Ed Engl 49:3904–3908

Wasmer C, Lange A, Van Melckebeke H, Siemer AB, Riek R, Meier BH (2008) Amyloid fibrils of
the HET-s(218-289) prion form a beta solenoid with a triangular hydrophobic core. Science
319:1523–1526

Wiltzius JJ, Sievers SA, Sawaya MR, Cascio D, Popov D, Riekel C, Eisenberg D (2008) Atomic
structure of the cross-b spine of islet amyloid polypeptide (amylin). Protein Sci 17:1467–1474

Wiltzius JJ, Landau M, Nelson R, Sawaya MR, Apostol MI, Goldschmidt L, Soriaga AB, Cascio
D, Rajashankar K, Eisenberg D (2009) Molecular mechanisms for protein-encoded inheritance. Nat Struct Mol Biol 16:973–978

Winner B, Jappelli R, Maji SK, Desplats PA, Boyer L, Aigner S, Hetzer C, Loher T, Vilar M,
Campioni S, Tzitzilonis C, Soragni A, Jessberger S, Mira H, Consiglio A, Pham E, Masliah E,
Gage FH, Riek R (2011) In vivo demonstration that {alpha}-synuclein oligomers are toxic.
Proc Natl Acad Sci USA 108:4194–4199

Seeding and Cross-seeding in Amyloid Diseases

Per Westermark and Gunilla T. Westermark

Abstract Seeding is a key phenomenon in all forms of amyloid and is most likely the mechanism by which amyloid deposits spread in a tissue and, in the case of systemic amyloidosis, from one organ to another. In experimental models of amyloid A (AA) amyloidosis, in which the fibril protein is derived from the acute phase reactant protein serum amyloid A (SAA), the disease can be transmitted from an amyloidotic animal to a susceptible recipient by a prion-like mechanism, and the transmitting agent has been identified as the amyloid fibril itself. AA amyloid fibrils from other species can also transfer the disease to mice, but with varying efficacy. In amyloid fibrils are protein monomers aggregated in a cross β-sheet conformation, and the molecular organization of amyloid fibrils is principally the same irrespective of biochemical type. However, fibril formation is specific, with subtle interactions between the β-strands making seeding specific, and even small amino acid sequence differences can block propagation. Nevertheless, cross-seeding does occur in vitro, and we have shown that fibrils of diverse nature, both natural and synthetic, can accelerate development of experimental AA amyloidosis. AA amyloidosis is common in many mammalian and avian species and can be present in the human food chain; such material can transfer amyloidosis experimentally to mice. Consequently, there are in our environment a number of components that may have the ability to induce amyloidosis in susceptible humans. It remains to be shown whether such mechanisms are limited to AA amyloidosis or can also operate in other protein aggregation disorders.

P. Westermark (✉)
Department of Immunology, Genetics and Pathology, Uppsala University,
751 85 Uppsala, Sweden
e-mail: Per.Westermark@igp.uu.se

G.T. Westermark
Department of Medical Cell Biology, Uppsala University, 75123 Uppsala, Sweden

M. Jucker and Y. Christen (eds.), *Proteopathic Seeds and Neurodegenerative Diseases*, 47
Research and Perspectives in Alzheimer's Disease,
DOI 10.1007/978-3-642-35491-5_4, © Springer-Verlag Berlin Heidelberg 2013

Amyloid and Amyloid Fibrils

Amyloid has long been described as an extracellular, inert substance, recognizable through particular staining properties, such as its affinity for the cotton dye Congo red and a green birefringence after this staining. Its fibrillar ultrastructure was described in 1959 (Cohen and Calkins 1959) but, in spite of its similar microscopic and ultrastructural appearance, it was soon understood that amyloid was not of uniform biochemical composition (Benditt and Eriksen 1962, 1971). Amino acid sequence analyses of purified proteins from clinically different forms of amyloid deposits revealed first a monoclonal immunoglobulin light chain origin (Glenner et al. 1971b) followed by protein amyloid A (AA) (Benditt et al. 1971), calcitonin (Sletten et al. 1976) and transthyretin (TTR; Costa et al. 1978). The number of known amyloid forms has gradually increased, and almost 30 different human proteins have now been identified as major components of amyloid fibrils (Sipe et al. 2012). The amyloid substance contains, in addition to the main fibrillar protein, other components, most uniformly serum amyloid P-component (SAP) and heparan sulfate proteoglycan (HSPG). The importance of these components is not fully understood, but SAP may help to stabilize the fibril and hinder degradation (Tennent et al. 1995) and heparan sulfate has been shown to support fibrillogenesis by increasing the β-sheet content of several amyloid fibril precursors, including serum amyloid A (SAA; McCubbin et al. 1988). Both SAP and HSPG are bound to amyloid fibrils by specific interactions.

Characteristic of all amyloid fibrils is that they are composed of one protein that differs between amyloid forms. The fibril backbone is a cross β-sheet structure with the sheets parallel to the fibril direction. Monomers are kept together by hydrogen bonds, but other forces, including van der Waal and electrostatic forces, are important. While it was originally thought that the whole amyloid protein had this structure (Eanes and Glenner 1968), it is now realized that the backbone may constitute only part of the protein (Wiltzius et al. 2009a). The Eisenberg group has shown that short segments of amyloid fibril proteins can be complementary and the side chains interact to form what they call 'steric zippers' (Sawaya et al. 2007; Wiltzius et al. 2009b). The zippers seem to be very specific and not the same in different amyloid proteins.

Fibril Formation

Much basic knowledge of the development of amyloid has come from in vitro experiments with model proteins (Glenner et al. 1971a, 1972; Jarrett and Lansbury 1993; Kelly and Lansbury 1994; Lai et al. 1996; Rochet and Lansbury 2000). Formation of amyloid fibrils in vitro is a nucleation-dependent phenomenon that has three distinct phases (Jarrett and Lansbury 1993). The first phase, which is energetically unfavorable, includes formation of a nucleus (Jarrett and Lansbury 1993), which serves as a site at which natively folded monomers are believed to misfold and aggregate into the first fibril. Since addition of new monomers takes place at the fibril ends, breakage of the fibril will result in further sites for amyloid growth (Moreno-Gonzales and Soto 2011). These events create the second, elongation phase. Finally, when the substrate concentration goes down, there is a steady state between fibril growth and degradation, constituting the third phase. Formation of the nucleus is the rate-limiting step and can take considerable time, known as the lag phase. However, seeding the protein solution with preformed fibrils eliminates this lag phase since fibril growth can start immediately on the added fibrils' ends.

Serum Amyloid A and AA Amyloidosis

The fibril protein in AA amyloidosis is protein AA, which is derived from its precursor serum amyloid A (SAA) by removal of a C-terminal fragment. Human SAA is a 104 amino acid protein expressed by three different genes. Two of these genes code for SAA1 and SAA2, respectively, and these gene products are acute phase reactant proteins, produced almost exclusively by the liver (Husby et al. 1994). Acute phase SAA production is under control of several cytokines, particularly IL-1, IL-6 and tumor necrosis factor (TNF; Uhlar and Whitehead 1999). In plasma, SAA is almost completely bound to high density lipoprotein (HDL) particles. The function of SAA is not well understood but there is strong evidence that it is important for the transport of cholesterol from injured tissue; however, a number of other effects have been described (Kisilevsky and Manley 2012). The plasma concentration of SAA is normally very low but can increase a thousand-fold upon inflammatory condition (Kisilevsky and Manley 2012). Some chronic inflammatory disorders, both infectious and non-infectious, may cause high plasma SAA levels for a very long time. Tuberculosis was once the most common cause of AA amyloidosis but has now been replaced by rheumatic diseases, particularly rheumatoid arthritis. A persistently high plasma concentration of SAA is a prerequisite for the development of AA amyloidosis, but since only a minority of these individuals develops amyloidosis, additional, yet unknown, factors must be important. These may be genetic or environmental. There are several isoforms of SAA 1 and 2; some are over-represented in amyloidosis, but this cannot be the sole explanation (Baba et al. 1995; Booth et al. 1998; Moriguchi et al. 2001). Other, unknown genetic factors may exist. No certain environmental factors have been identified but, given experimental results with mouse models (see below), seeding with a variety of possible materials is a realistic but so far only hypothetic possibility.

Seeding in vivo in 'peripheral' amyloid forms has been studied extensively in two murine models of systemic amyloidoses of AA and apolipoprotein A-II (AApoAII) nature. AA amyloidosis resembling the human form occurs in many mammalian and avian species, both in the wild and in captivity. Experimentally induced AA amyloidosis is the oldest animal model of any form of amyloidosis and has been used since the beginning of the twentieth century, first in rabbits (Dantchakow 1907) but later particularly in mice.

AA amyloidosis can easily be induced in many strains of mice by the induction of a chronic inflammatory condition. Originally, animals were infected with bacteria, e.g., streptococci, but non-infectious models soon became more common (Jaffé 1926). Typical methods have been induction by daily subcutaneous injections of casein (Druet and Janigan 1966) or by the creation of a sterile abscess with silver nitrate (Kisilevsky et al. 1979). Depending on several conditions, including mouse strain, a systemic amyloidosis develops after several weeks. Typically, the first amyloid is to be found perifollicularly in the spleen, but deposits in liver soon follows, as well as amyloid in other organs. The distribution of amyloid is similar to that of human AA amyloidosis. In 1966, spleen cells from an amyloidotic or pre-amyloidotic mouse were found to transfer the disease to a sensitized recipient (Hardt and Ranløv 1976; Ranløv 1967; Werdelin and Ranløv 1966). The ability to transfer was verified by studies of other groups (Axelrad et al. 1982; Janigan and Druet 1968; Kisilevsky et al. 1979; Varga et al. 1986; Willerson et al. 1969). An interesting study was reported by Benditt (1976). In both his own studies with Peking ducks and anecdotal reports with the same species and with mice, he showed evidence of a factor that was transferred between animals in captivity and that induced or enhanced the development of AA amyloidosis. Benditt seems to have been the first, and for long time the only one, to associate the phenomenon of transfer amyloid with the experimental transmissibility of kuru (Gajdusek et al. 1966).

Very similar results have been obtained with the AApoA-II type of murine amyloidosis (Fu et al. 2004; Korenaga et al. 2004; Naiki et al. 1991). Trials with this model to inactivate the transmission by diverse means showed remarkable resemblance to prions; only strong physical and chemical methods completely inactivated the material (Zhang et al. 2006).

Amyloid Enhancing Factor Is Aggregated Protein

The nature of the agent, called amyloid enhancing factor (AEF), causing the transmissibility of AA amyloidosis was long a puzzle. Diverse factors including virus and nuclear factors were suggested. However, a fibril-associated factor was also discussed, possibly in some unknown way acting as a template for further fibril formation (Niewold et al. 1986). In 2002, we reported strong evidence that AEF was, in reality, the AA fibril itself and that misfolded and aggregated AA molecules speed up development of amyloidosis in susceptible animals by a prion-like mechanism (Lundmark et al. 2002). In a biochemical analysis of the active material, only SAA-derived molecules were identified (Lundmark et al. 2002), although additional, minor components, such as SAP and HSPG, must have been present. The analyzed AEF preparation, which essentially consisted only of fibrils extracted from mice with severe amyloidosis, was surprisingly effective and could be diluted serially many times without loss of potency (Lundmark et al. 2002). The view that fibrils themselves constituted AEF was further strengthened by the finding that amyloid-like fibrils made from short, synthetic peptides corresponding to parts of amyloid fibril proteins also exerted an amyloid-enhancing effect, although less dramatic (Ganowiak et al. 1994; Johan et al. 1998). A puzzling observation of ours is that fibrils made from recombinant mouse AA are not as efficient as AEF. This difference resembles that with transmissible spongiform encephalopathies, where it has been difficult to establish infectious particles from recombinant prion protein (Colby and Prusiner 2011). The reasons are unclear, but subtle differences in fibril conformation may be of importance, as well as effects of additional factors. In the case of AA amyloidosis, HSPG may be such a factor.

Experimental acceleration of AA amyloidosis with the aid of AEF preparations is not restricted to mice. It has been shown with several other rodent species, including hamster and guinea pig (Niewold et al. 1987) (although guinea pig is perhaps not a rodent; D'Erchia et al. 1996), as well as animals of other orders such as mink (Sørby et al. 2008). It is therefore probable that the mechanism would be the same in humans.

Murine AEF of AA amyloid nature is potent not only when given intravenously but also via the oral route. When animals were given AEF in drinking water before an inflammation was induced, the lag phase was abolished (Lundmark et al. 2002). Administration of AEF in the nostril at induction of inflammation had the same effect. Oral transmission of AA amyloidosis to mice with fibrils from different species has been reported (Cui et al. 2002), and the oral route for transmission of AA amyloidosis in cheetahs in captivity has also been reported (Zhang et al. 2008). Not only is AA amyloidosis transmissible orally but the biochemically different AapoAII amyloidoses in mice (Xing et al. 2001) and probably in cheetah (Zhang et al. 2008) are also 'infective' in this way, showing that oral transmission is not restricted to rodents.

Experimental transmission of systemic amyloidosis has not been shown for other systemic amyloidoses. There are no good models of AL amyloidosis, and ATTR amyloidosis, for which there are some mouse models, is not seedable since the rate-limiting step for formation of ATTR amyloidosis (for nomenclature, see Sipe et al. 2010) is dissociation of the TTR tetramer (Hurshman et al. 2004). However, as soon as monomers have been generated, TTR should also be seedable and seeding is most probably the way in which ATTR deposits are spread throughout the body. There is indirect evidence for this statement. First, amyloid deposits in familial ATTR amyloidosis often continue to develop even after the source of mutant TTR has been removed, and the amyloid can finally consist of almost 100 % wild-type protein (Ihse et al. 2011). Second, familial ATTR amyloidosis is often treated with liver transplantation since the liver is the main producer of plasma TTR. Replacement with a liver expressing only wild-type protein is consequently a form of gene therapy. Since there is shortage of livers, the otherwise healthy organ from the patient with amyloidosis is usually used for a second receiver, often a patient with liver malignancy or end-stage liver disease. In this so-called domino transplantation, the receiver of a transplant from the amyloidosis patient has in a number of cases developed typical ATTR amyloidosis (Stangou et al. 2005).

On the other hand, there are several examples of seeding and transmission in localized forms of amyloid and amyloid-like aggregates, particularly in the brain (Dunning et al. 2012; Eisele et al. 2009; Jucker and Walker 2011; Morales et al. 2010). These include prion disorders, Alzheimer's disease and Parkinson's disease, among others (Brundin et al. 2010). Although most studies on these mechanisms have been performed on central nervous system diseases, there is no reason to believe that amyloid outside the CNS behaves differently (Westermark and Westermark 2010).

Cross-seeding

Seeding is a key phenomenon in the formation of all forms of amyloid deposits and is believed to be the mechanism by which amyloid spreads between cells and in tissues and, in systemic amyloidoses, from one organ to another. Although seeding generally is quite specific and even very small differences in amino acid sequence can abolish induction of new fibrils, cross-seeding can occur. Thus cross-seeding with fibrils from other animal species has been shown with the mouse AA amyloid model, but with varying efficacy. All amyloid fibrils, irrespective of biochemical origin, have a cross β-sheet structure, and amyloid-like fibrils made in vitro from short synthetic peptides corresponding to parts of amyloid fibril proteins of diverse origin have also been found to induce AA amyloidosis (Johan et al. 1998).

Cross-seeding means seeding with aggregates of a different protein nature. From this definition, it can be questioned whether seeding a protein with fibrils of the same nature but from another species should be called cross-seeding. The commonly described species barriers could have another background (Scott et al. 2005). Colocalization of polyglutamine inclusions and α-synuclein immunoreactivity has been described that could be an example of 'true' cross-seeding (Charles et al. 2000). Cross-seeding has been induced experimentally with two biochemical forms of systemic amyloidosis: AA and AApoAII (Yan et al. 2007). The idea that AEF in transmission of AA amyloidosis depended on a seeding mechanism by β-pleated sheet fibrils led to experiments where fibrils made in vitro from short synthetic peptides were used for induction of amyloidosis in susceptible mice (Figs. 1 and 2; Ganowiak et al. 1994; Johan et al. 1998). The synthetic peptides corresponded to segments of other amyloid fibril proteins (TTR and islet amyloid polypeptide) with no amino acid sequence identity but had, nevertheless, an amyloid accelerating effect. Researchers working with the apolipoprotein A-II model obtained similar results (Fu et al. 2004). Consequently, fibrils with amyloid-type conformation but of different biochemical nature can template misfolding of SAA molecules.

Mechanisms of Cross-seeding

The molecular mechanisms behind cross-seeding are not understood. In the case of close similarity between the seed and the target molecule, e.g., prion proteins of different species, the effect is probably the same as in seeding with identical molecules. However, seeding with amyloid-like fibrils made from short synthetic peptides from other amyloid proteins (Johan et al. 1998) or with fibrils of a completely

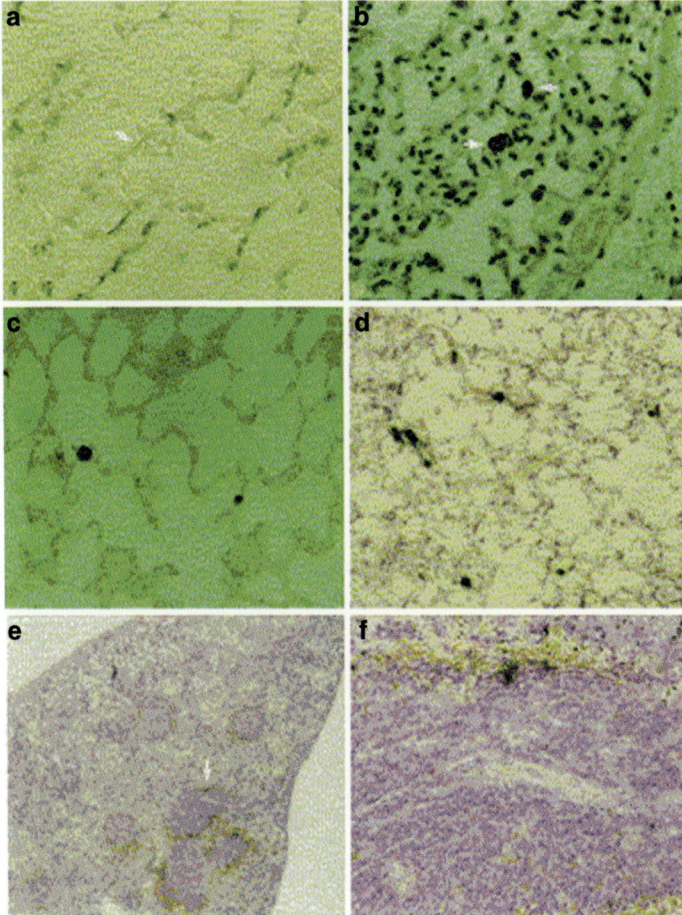

Fig. 1 Example of cross-seeding. Fibrils were made in vitro from a peptide corresponding to amino acid residues 115–124 of human transthyretin and injected intravenously into inflamed mice with a high expression of SAA. Small amyloid thrombi got stuck in lung capillaries, shown in (**a**) in *Congo red*. These amyloid thrombi were strongly immunoreactive both for protein AA (**b**) and for injected fibril peptide (**c**). (**d**) The thrombi shown by autoradiography, where the injected fibrils had been radioactively labeled. (**e**) A section of spleen and silver grains occur perifollicularly at the amyloid deposits. (**f**) Close-up of e at the arrow (figure from Johan et al. 1998)

different nature, e.g., bacterial curli or Sup35 from Saccharomyces (Lundmark et al. 2005), may work by other interactions. It seems likely that the β-sheet structure of the seed has some general effect on the seeded material, inducing a misfolding and a new seed that can proceed to an amyloid fibril.

In vitro experiments with pure proteins have been informative but have also produced results that are difficult to interpret. The two peptides Aβ and islet amyloid polypeptide (IAPP), the main fibril components in cerebral amyloid in Alzheimer's disease and in the islets of Langerhans in type 2 diabetes, respectively,

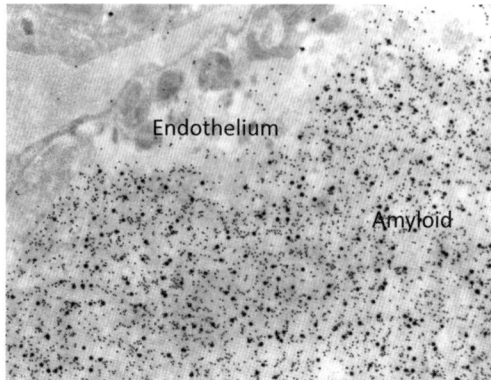

Fig. 2 The same material as in Fig. 1, shown electron microscopically after double labeling with antibodies against protein AA (visualized with 10-nm gold particles) and transthyretin 115–124 (visualized with 20-nm gold particles). It is obvious that the amyloid fibrils are equally strongly labeled with the two antibodies

have sequential similarities. Aβ fibrils seeded IAPP efficiently but the opposite was not seen (O'Nuallian et al. 2004). Interestingly, very small changes in Aβ sequence altered the effect substantially.

In the case of Aβ and IAPP, the proteins have some sequence identity and a cross-seeding can seem reasonable. However, an AA protein in solution was seeded by fibrils made from medin and these two proteins were completely non-related and without amino acid sequence identity (Larsson et al. 2011). Since the interaction between β-strands to form the fibril spine seems very specific (Goldschmidt et al. 2010; Wiltzius et al. 2009a), it appears likely that, in the cross-seeding mechanism, a steric zipper interaction between the seed and the seeded protein is not necessarily formed. Perhaps some looser interaction between the seed's backbone and the seeded protein can take place, inducing β-sheet fibrils in the recipient. In that case, co-fibril formation would not occur. In fact, in vivo co-deposition of proteins in amyloid fibrils usually does not take place and, when two different amyloid deposits are found in the same individual, the deposits are generally not mixed (Bergström et al. 2004; Larsson et al. 2011).

Is There a Link Between Different Amyloid Diseases?

As stated above, the amyloid deposits found in vivo do not usually consist of a mixture of fibril proteins, but this fact does not necessarily mean that one biochemical type of amyloid is incapable of inducing fibril formation from another by seeding. There is a link between type 2 diabetes and Alzheimer's disease (Han and Li 2010), and this association raises the intriguing question whether there may be a connection between amyloid deposits in the two disorders. We found experimentally that intravenous injection of Aβ fibrils strongly enhances the development of islet amyloid in a transgenic mouse strain, expressing human IAPP. An increased risk of type 2 diabetes in Alzheimer's disease has been suggested (Janson et al. 2004). However, others have found an increased risk of Alzheimer pathology in type 2 diabetes and suggested that this may partially depend on hyperinsulinemia, often seen in type 2 diabetes (Matsuzaki et al. 2010).

In the examples of senile systemic amyloidosis (derived from wild-type TTR) and apolipoprotein A-IV (AApoA-IV) amyloidosis, amyloid deposits of both kinds were found side by side at the same sites (Bergström et al. 2004). Although the deposits were not mixed, a cross-seeding mechanism can be suspected, particularly since apolipoprotein A-IV-derived amyloidosis is exceedingly rare. A similar finding was obtained by studies of aortic sections of subjects with AA amyloidosis. Amyloid occurs almost constantly in the media of the thoracic aorta of elderly humans (Mucchiano et al. 1992; Schwartz 1970). The fibrils consist of the 50 amino-acid-residue protein medin, which is a fragment of the protein lactadherin, expressed by aortic smooth muscle cells (Häggqvist et al. 1999). In the aortic wall of the studied patients, deposits of both AA and medin (AMed) amyloid were found but usually not together (Larsson et al. 2011). However, it is still possible that one of the amyloids seeded the other one but that the kinetics was different.

Seeding and Cross-seeding of Human Amyloidosis by Environmental Factors

As stated above, human amyloid forms are almost never biochemically mixed, but this does not rule out the possibility that one amyloid form can induce another one in vivo. Amyloid and amyloid-like materials are present in our environment, including our food. AA amyloid is common both in captive and wild animals and is more common than believed in human food (Solomon et al. 2007; Yoshida et al. 2009). Such amyloid has been shown to seed AA amyloidosis in susceptible mice (Solomon et al. 2007) and would likely do so also in humans with a persistently high plasma concentration of SAA. However, transmission to humans has not been shown so far and it would be difficult to detect.

Many naturally occurring fibers have amyloid-type molecular organization. Such fibers include, for example, silk, bacterial curli and some fungal proteins (Lundmark et al. 2005). These materials also have the ability to accelerate development of murine AA amyloidosis (Kisilevsky et al. 1999; Lundmark et al. 2005). Some polypeptide-based materials in nanotechnology also utilize amyloid fibril-like aggregation (Ellis-Behnke et al. 2006), and such fibrils seed murine AA amyloidosis with varying efficacy (Westermark et al. 2009). Consequently, might the introduction of amyloid-like structures mean new possible risk factors for protein aggregation diseases?

Experiments in vitro with β-amyloid and prion protein have shown that the morphology, and perhaps behavior, of fibrils depends on the seed and that different 'strains' can be obtained (Makarava and Baskakov 2008; Petkova et al. 2005). If cross-seeding is important for the development of human amyloid diseases, one may question whether different variants of disease will be the result depending on the nature of cross-seeding material. This question should be pertinent both to localized amyloids, e.g., of Aβ nature, and to systemic forms of amyloidosis. The mechanism could provide a possible cause of the morphologic variations seen in all types of amyloid and which have not been explained so far.

Conclusion

From the compiled literature concerning several animal models of amyloid diseases, both localized and systemic, it is absolutely clear that transmission of the disease can be performed experimentally by a prion-like mechanism. There is also strong evidence that such a mechanism is operative in animals in captivity and therefore likely in nature, similar to what is seen with the prion disorder chronic wasting disease in deer (Haley et al. 2009; Mathiason et al. 2009). There is no reason to believe that humans are resistant to the principle. Whether cross-seeding occurs naturally is still unknown but offers a challenging mechanism of induction of all kinds of amyloid-associated diseases.

Acknowledgments Supported by the Swedish Research Council, the Swedish Diabetes Association and FAMY, FAMY Norrbotten and Amyl.

References

Axelrad MA, Kisilevsky R, Willmer J, Chen SJ, Skinner M (1982) Further characterization of amyloid-enhancing factor. Lab Invest 47:139–146

Baba S, Masago SA, Takahashi T, Kasama T, Sugimura H, Tsugane S, Tsutsui Y, Shirasawa H (1995) A novel allelic variant of serum amyloid A, SAA1g: genomic evidence, evolution, frequency, and implication as a risk factor for reactive systemic AA-amyloidosis. Hum Mol Genet 4:1083–1087

Benditt EP (1976) The structure of amyloid protein AA and evidence for a transmissible factor in the origin of amyloidosis. In: Wegelius O, Pasternack A (eds) Amyloidosis. Academic, London, pp 323–331

Benditt EP, Eriksen N (1962) Chemical similarity among amyloid substances associated with long standing inflammation. Lab Invest 26:615–625

Benditt EP, Eriksen N (1971) Chemical classes of amyloid substance. Am J Pathol 65:231–252

Benditt EP, Eriksen N, Hermodson MA, Ericsson LH (1971) The major proteins of human and monkey amyloid substance: common properties including unusual N-terminal amino acid sequences. FEBS Lett 19:169–173

Bergström J, Murphy CL, Weiss DT, Solomon A, Sletten K, Hellman U, Westermark P (2004) Two different types of amyloid deposits – apolipoprotein A-IV and transthyretin – in a patient with systemic amyloidosis. Lab Invest 84:981–988

Booth DR, Booth SE, Gillmore JD, Hawkins PN, Pepys MB (1998) SAA1 alleles as risk factors in reactive systemic AA amyloidosis. Amyloid 5:262–265

Brundin P, Melki R, Kopito R (2010) Prion-like transmission of protein aggregates in neurodegenerative disease. Mol Cell Biol 11:301–307

Charles V, Mezeyb E, Reddya PH, Dehejiaa A, Younga TA, Polymeropoulos MH, Brownstein MJ, Tagle DA (2000) Alpha-synuclein immunoreactivity of huntingtin polyglutamine aggregates in striatum and cortex of Huntington's disease patients and transgenic mouse models. Neurosci Lett 289:29–32

Cohen AS, Calkins E (1959) Electron microscopic observations on a fibrous component in amyloid of diverse origins. Nature 183:1202–1203

Colby DW, Prusiner SB (2011) De novo generation of prion strains. Nat Rev Microbiol 9:771–777

Costa PP, Figueira AS, Bravo FR (1978) Amyloid fibril protein related to prealbumin in familial amyloidotic polyneuropathy. Proc Natl Acad Sci USA 75:4499–4503

Cui D, Kawano H, Takahashi M, Hoshii Y, Setoguchi M, Gondo T, Ishihara T (2002) Acceleration of murine AA amyloidosis by oral administration of amyloid fibrils extracted from different species. Pathol Int 52:40–45

D'Erchia AM, Gissi C, Pesole G, Saccone C, Arnason U (1996) The guinea-pig is not a rodent. Nature 381:597–600

Dantchakow W (1907) Über die entwicklung und resorption experimentell erzeugter amyloidsubstanz in der speicheldrüsen von kaninchen. Virchows Arch 187:1–34

Druet RL, Janigan DT (1966) Experimental amyloidosis. Rates of induction, lymphocyte depletion and thymic atrophy. Am J Pathol 49:911–929

Dunning CJ, Reyes JF, Steiner JA, Brundin P (2012) Can Parkinson's disease pathology be propagated from one neuron to another? Prog Neurobiol 97:205–219

Eanes ED, Glenner GG (1968) X-ray diffraction studies on amyloid filaments. J Histochem Cytochem 16:673–677

Eisele YS, Bolmont T, Heikenwalder M, Langer F, Jacobson LH, Yan ZX, Roth K, Aguzzi A, Staufenbiel M, Walker LC, Jucker M (2009) Induction of cerebral β-amyloidosis: intracerebral versus systemic Aβ inoculation. Proc Natl Acad Sci USA 106:12926–12931

Ellis-Behnke RG, Liang YX, You SW, Tay DK, Zhang S, So KF, Schneider GE (2006) Nano neuro knitting: peptide nanofiber scaffold for brain repair and axon regeneration with functional return of vision. Proc Natl Acad Sci USA 103:5054–5059

Fu X, Korenaga T, Fu L, Xing Y, Guo Z, Matsushita T, Hosokawa M, Naiki H, Baba S, Kawata Y, Ikeda S, Ishihara T, Mori M, Higuchi K (2004) Induction of AApoAII amyloidosis by various heterogeneous amyloid fibrils. FEBS Lett 563:179–184

Gajdusek DC, Gibbs CJ Jr, Alpers M (1966) Experimental transmission of a Kuru-like syndrome to chimpanzees. Nature 209:794–796

Ganowiak K, Hultman P, Engström U, Gustavsson Å, Westermark P (1994) Fibrils from synthetic amyloid-related peptides enhance development of experimental AA-amyloidosis in mice. Biochem Biophys Res Commun 199:306–312

Glenner GG, Ein D, Eanes ED, Bladen HA, Terry W, Page D (1971a) The creation of "amyloid" fibrils from Bence Jones proteins in vitro. Science 174:712–714

Glenner GG, Terry W, Harada M, Isersky C, Page D (1971b) Amyloid fibril proteins: proof of homology with immunoglobulin light chains by sequence analysis. Science 172:1150–1151

Glenner GG, Eanes ED, Page DL (1972) The relation of the properties of Congo red-stained amyloid fibrils to the β-conformation. J Histochem Cytochem 20:821–826

Goldschmidt L, Teng PK, Riek R, Eisenberg D (2010) Identifying the amylome, proteins capable of forming amyloid-like fibrils. Proc Natl Acad Sci USA 107:3487–3492

Häggqvist B, Näslund J, Sletten K, Westermark GT, Mucchiano G, Tjernberg LO, Nordstedt C, Engström U, Westermark P (1999) Medin: an integral fragment of aortic smooth muscle cell-produced lactadherin forms the most common human amyloid. Proc Natl Acad Sci USA 96:8669–8674

Haley NJ, Seelig DM, Zabel MD, Telling GC, Hoover EA (2009) Detection of CWD prions in urine and saliva of deer by transgenic mouse bioassay. PLoS One 4:e4848

Han W, Li C (2010) Linking type 2 diabetes and Alzheimer's disease. Proc Natl Acad Sci USA 107:6557–6558

Hardt F, Ranløv PJ (1976) Transfer amyloidosis. Int Rev Exp Pathol 16:273–334

Hurshman AR, White JT, Powers ET, Kelly JW (2004) Transthyretin aggregation under partially denaturing conditions is a downhill polymerization. Biochemistry 43:7365–7381

Husby G, Marhaug G, Dowton B, Sletten K, Sipe JD (1994) Serum amyloid A (SAA): biochemistry, genetics and the pathogenesis of AA amyloidosis. Amyloid 1:119–137

Ihse E, Suhr OB, Hellman U, Westermark P (2011) Variation in amount of wild-type transthyretin in different fibril and tissue types in ATTR amyloidosis. J Mol Med 89:171–180

Jaffé RH (1926) Experimental amyloidosis in mice. Effect of different forms of diet. Arch Pathol Lab Med 2:149

Janigan DT, Druet RL (1968) Experimental murine amyloidosis in x-irradiated recipients of spleen homogenates or serum from sensitized donors. Am J Pathol 52:381–390

Janson J, Laedtke T, Parisi JE, O'Brien P, Petersen RC, Butler PC (2004) Increased risk of type 2 diabetes in Alzheimer disease. Diabetes 53:474–481

Jarrett JT, Lansbury PT (1993) Seeding "one-dimensional crystallization" of amyloid: a pathogenic mechanism in Alzheimer's disease and scrapie? Cell 73:1055–1058

Johan K, Westermark GT, Engström U, Gustavsson Å, Hultman P, Westermark P (1998) Acceleration of AA-amyloidosis by amyloid-like synthetic fibrils. Proc Natl Acad Sci USA 95:2558–2563

Jucker M, Walker LC (2011) Pathogenic protein seeding in Alzheimer disease and other neurodegenerative disorders. Ann Neurol 70:532–540

Kelly JW, Lansbury PTJ (1994) A chemical approach to elucidate the mechanism of transthyretin and β-protein amyloid fibril formation. Amyloid 1:186–205

Kisilevsky R, Manley PN (2012) Acute phase serum amyloid A: perspectives on its physiological and pathological roles. Amyloid 19:5–14

Kisilevsky R, Benson MD, Axelrad MA, Boudreau L (1979) The effect of a liver protein synthesis inhibitor on plasma SAA levels in a model of accelerated amyloid deposition. Lab Invest 41:206–210

Kisilevsky R, Lemieux L, Boudreau L, Yang DS, Fraser P (1999) New clothes for amyloid enhancing factor (AEF): silk as AEF. Amyloid 6:98–106

Korenaga T, Fu X, Xing Y, Matsusita T, Kuramoto K, Syumiya S, Hasegawa K, Niaiki H, Ueno M, Ishihara T, Hosokawa M, Mori M, Higuchi K (2004) Tissue distribution, biochemical properties, and transmission of mouse type A AApoAII amyloid fibrils. Am J Pathol 164:1597–1606

Lai Z, Colón W, Kelly JW (1996) The acid-mediated denaturation pathway of transthyretin yields a conformational intermediate that can self-assemble into amyloid. Biochemistry 35:6470–6482

Larsson A, Malmström S, Westermark P (2011) Signs of cross-seeding: aortic medin amyloid as a trigger for protein AA deposition. Amyloid 18:229–234

Lundmark K, Westermark GT, Nyström S, Murphy CL, Solomon A, Westermark P (2002) Transmissibility of systemic amyloidosis by a prion-like mechanism. Proc Natl Acad Sci USA 99:6979–6984

Lundmark K, Westermark GT, Olsén A, Westermark P (2005) Protein fibrils in nature can enhance AA amyloidosis in mice: cross-seeding as a disease mechanism. Proc Natl Acad Sci USA 102:6098–6102

Makarava N, Baskakov IV (2008) The same primary structure of the prion protein yields two distinct self-propagating states. J Biol Chem 283:15988–15996

Mathiason CK, Hays SA, Powers J, Hayes-Klug J, Langenberg J, Dahmes SJ, Osborn DA, Miller KV, Warren RJ, Mason GL, Hoover EA (2009) Infectious prions in pre-clinical deer and transmission of chronic wasting disease solely by environmental exposure. PLoS One 4:e5916

Matsuzaki T, Sasaki K, Tanizaki Y, Hata J, Fujimi K, Matsui Y, Sekita A, Suzuki SO, Kanba S, Kiyohara Y, Iwaki T (2010) Insulin resistance is associated with the pathology of Alzheimer disease: the Hisayama study. Neurology 75:764–770

McCubbin WD, Kay CM, Narindrasorasak S, Kisilevsky R (1988) Circular-dichroism studies on two murine serum amyloid A proteins. Biochem J 256:775–783

Morales R, Estrada LD, Diaz-Espinoza R, Morales-Scheihing D, Jara MC, Castilla J, Soto C (2010) Molecular cross talk between misfolded proteins in animal models of Alzheimer's and prion diseases. J Neurosci 30:4528–4535

Moreno-Gonzales I, Soto C (2011) Misfolded protein aggregates: mechanisms, structures and potential for disease transmission. Semin Cell Dev Biol 22:482–487

Moriguchi M, Terai C, Kaneko H, Koseki Y, Kajiyama H, Uesato M, Inada S, Kamatani N (2001) A novel single-nucleotide polymorphism at the 5'-flanking region of SAA1 associated with risk of type AA amyloidosis secondary to rheumatoid arthritis. Arthritis Rheum 44:1266–1272

Mucchiano G, Cornwell GGI, Westermark P (1992) Senile aortic amyloid. Evidence of two distinct forms of localized deposits. Am J Pathol 140:871–877

Naiki H, Higuchi K, Nakakuki K, Takeda T (1991) Kinetic analysis of amyloid fibril polymerization in vitro. Lab Invest 65:104–110

Niewold TA, van Andel ACJ, Hol PR, Gruys E (1986) Fibril derived amyloid enhancing factor (FAEF) in hamster: evidence for a close relationship between AEF and AA-amyloid fibrils. In: Marrink J, van Rijswijk MH (eds) Amyloidosis. Martinus Nijhoff, Dordrecht, pp 177–182

Niewold TA, Hol PR, van Andel AC, Lutz ET, Gruys E (1987) Enhancement of amyloid induction by amyloid fibril fragments in hamster. Lab Invest 56:544–549

O'Nuallian B, Williams AD, Westermark P, Wetzel R (2004) Seeding specificity in amyloid growth induced by heterologous fibrils. J Biol Chem 279:17490–17499

Petkova AT, Leapman RD, Guo Z, Yau W-M, Mattson MP, Tycko R (2005) Self-propagating, molecular-level polymorphism in Alzheimer's β-amyloid fibrils. Science 307:262–265

Ranløv P (1967) The adoptive transfer of experimental mouse amyloidosis by intravenous injections of spleen cell extracts from casein-treated syngeneic donor mice. Acta Pathol Microbiol Scand 70:321–335

Rochet J-C, Lansbury PTJ (2000) Amyloid fibrillogenesis: themes and variations. Curr Opin Struct Biol 10:60–68

Sawaya MR, Sambashivan S, Nelson R, Ivanova MI, Sievers SA, Apostol MI, Thompson MJ, Balbirnie M, Wiltzius JJ, McFarlane HT, Madsen AØ, Riekel C, Eisenberg D (2007) Atomic structures of amyloid cross-beta spines reveal varied steric zippers. Nature 447:453–457

Schwartz P (1970) Amyloidosis, cause and manifestation of senile deterioration. C.C. Thomas, Springfield, IL

Scott MR, Peretz D, Nguyen H, Dearmond SJ, Prusiner SB (2005) Transmission barriers for bovine, ovine, and human prions in transgenic mice. J Virol 79:5259–5271

Sipe JD, Benson MD, Buxbaum JN, Ikeda S, Merlini G, Saraiva MJ, Westermark P (2012) Amyloid fibril protein nomenclature: 2012 recommendations of the nomenclature committee of the Intl Soc Amyloidosis. Amyloid 19:167–170

Sletten K, Westermark P, Natvig JB (1976) Characterization of amyloid fibril proteins from medullary carcinoma of the thyroid. J Exp Med 143:993–998

Solomon A, Richey T, Murphy CL, Weiss DT, Wall JS, Westermark GT, Westermark P (2007) Amyloidogenic potential of foie gras. Proc Natl Acad Sci USA 104:10998–11001

Sørby R, Espenes A, Landsverk T, Westermark G (2008) Rapid induction of experimental AA amyloidosis in mink by intravenous injection of amyloid enhancing factor. Amyloid 15:21–29

Stangou AJ, Heaton ND, Hawkins PN (2005) Transmission of systemic transthyretin amyloidosis by means of domino liver transplantation. N Engl J Med 352:2356

Tennent GA, Lovat LB, Pepys MB (1995) Serum amyloid P component prevents proteolysis of the amyloid fibrils of Alzheimer disease and systemic amyloidosis. Proc Natl Acad Sci USA 92:4299–4303

Uhlar CM, Whitehead AS (1999) Serum amyloid A, the major vertebrate acute-phase reactant. Eur J Biochem 265:501–523

Varga J, Flinn MSM, Shirahama T, Rodgers OG, Cohen AS (1986) The induction of accelerated murine amyloid with human splenic extract. Probable role of amyloid enhancing factor. Virchows Arch 51:177–185

Werdelin O, Ranløv P (1966) Amyloidosis in mice produced by transplantation of spleen cells from casein-treated mice. Acta Pathol Microbiol Scand 68:1–18

Westermark GT, Westermark P (2010) Prion-like aggregates: infectious agents in human disease. Trends Mol Med 16:501–507

Westermark P, Lundmark K, Westermark GT (2009) Fibrils from designed non-amyloid-related synthetic peptides induce AA-amyloidosis during inflammation in an animal model. PLoS One 4:e6041

Willerson JT, Gordon JK, Talal N, Barth WF (1969) Murine amyloid. II. Transfer of an amyloid-accelerating substance. Arthritis Rheum 12:232–240

Wiltzius JJ, Landau M, Nelson R, Sawaya MR, Apostol MI, Goldschmidt L, Soriaga AB, Cascio D, Rajashankar K, Eisenberg D (2009a) Molecular mechanisms for protein-encoded inheritance. Nat Struct Mol Biol 16:973–978

Wiltzius JJW, Sievers SA, Sawaya MR, Eisenberg D (2009b) Atomic structures of IAPP (amylin) fusions suggest a mechanism for fibrillation and the role of insulin in the process. Protein Sci 18:1521–1530

Xing Y, Nakamura A, Chiba T, Kogishi K, Matsushita T, Li F, Guo Z, Hosokawa M, Mori M, Higuchi K (2001) Transmission of mouse senile amyloidosis. Lab Invest 81:493–499

Yan J, Fu X, Ge F, Zhang B, Yao J, Zhang H, Qian J, Tomozawa H, Naiki H, Sawashita J, Mori M, Higuchi K (2007) Cross-seeding and cross-competition in mouse apolipoprotein A-II amyloid fibrils and protein A amyloid fibrils. Am J Pathol 171:172–180

Yoshida T, Zhang P, Fu X, Higuchi K, Ikeda S-I (2009) Slaughtered aged cattle might be one dietary source exhibiting amyloid enhancing factor activity. Amyloid 16:25–31

Zhang H, Sawashita J, Fu X, Korenaga T, Yan J, Mori M, Higuchi K (2006) Transmissibility of mouse AApoAII amyloid fibrils: inactivation by physical and chemical methods. FASEB J 20: E211–E221

Zhang B, Une Y, Fu X, Yan J, Ge F, Yao J, Sawashita J, Mori M, Tomozawa H, Kametani F, Higuchi K (2008) Fecal transmission of AA amyloidosis in the cheetah contributes to high incidence of disease. Proc Natl Acad Sci USA 105:7263–7268

The Prion-Like Aspect of Alzheimer Pathology

Sarah K. Fritschi, Bahareh Eftekharzadeh, Giusi Manfredi,
Tsuyoshi Hamaguchi, Götz Heilbronner, Amudha Nagarathinam,
Franziska Langer, Yvonne S. Eisele, Lary Walker,
and Mathias Jucker

Abstract Many neurodegenerative disorders are characterized by a predictable spatiotemporal progression of the aggregation of specific proteins in the brain. The most prevalent cerebral proteopathy is Alzheimer's disease (AD), in which aggregated amyloid-β peptide (Aβ) is deposited in the form of extracellular parenchymal plaques and vascular amyloid. Multiple lines of evidence indicate that β-amyloidosis can be exogenously induced by the application of brain extracts containing aggregated Aβ. The β-amyloid-inducing agent in the extract is likely Aβ itself in a conformation that cannot easily be mimicked with synthetic material. The induced Aβ lesions spread over time within and among brain regions, and they are dependent on the structural and biochemical nature of Aβ in the extract and on the characteristics of the host. We have found that bioactive Aβ seeds exist in both soluble and insoluble forms; some of them are sensitive to proteinase-K digestion and some are not. Observations of similar prion-like induction, spreading, and transmission of tau lesions, the second hallmark of AD pathology, and more recent observations of seeded α-synuclein lesions suggest that the concept of prion-like corruptive templating of proteins may also apply to intracellular lesions in neurodegenerative diseases. The clinical implications of these observations are not yet clear. The finding that the Aβ seeds are partly soluble suggests that such seeds in bodily fluids may have diagnostic value and also that they could represent a novel target for early therapeutic intervention. Furthermore, the possibility that mechanisms exist

S.K. Fritschi • B. Eftekharzadeh • G. Manfredi • T. Hamaguchi • G. Heilbronner •
A. Nagarathinam • F. Langer • Y.S. Eisele • M. Jucker (✉)
Department of Cellular Neurology, Hertie Institute for Clinical Brain Research, University of
Tübingen, Otfried-Mueller Strasse 27, 72076 Tübingen, Germany

DZNE, German Center for Neurodegenerative Diseases, 72076 Tübingen, Germany
e-mail: mathias.jucker@uni-tuebingen.de

L. Walker
Yerkes National Primate Research Center, Emory University, Atlanta, GA 30329, USA

Department of Neurology, Emory University, Atlanta, GA 30329, USA

M. Jucker and Y. Christen (eds.), *Proteopathic Seeds and Neurodegenerative Diseases*,
Research and Perspectives in Alzheimer's Disease,
DOI 10.1007/978-3-642-35491-5_5, © Springer-Verlag Berlin Heidelberg 2013

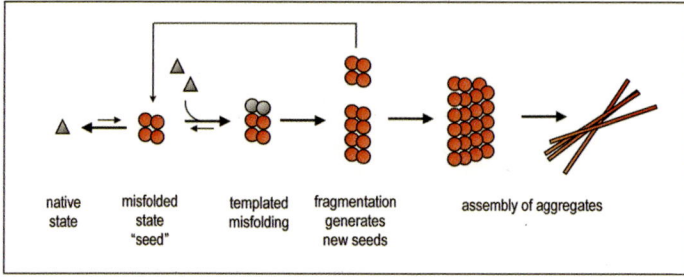

Fig. 1 Initiation of protein misfolding by a nucleation-dependent seeding process. This model (adapted from Harper and Lansbury 1997) proposes that the native and the misfolded states of the Aβ peptide are reversible. The native conformation is strongly favored. Below a critical concentration of the protein, no aggregation occurs, but when the protein concentration exceeds a critical level, the protein can form a corruptive seed that then expands the templated misfolding to other Aβ monomers. Addition of an exogenous seed eliminates the lag period and rapidly induces aggregation of the protein. Fragmentation of larger aggregates generates new seeds and therefore accelerates amyloid formation. In vivo, the progressive protein aggregation is accompanied by pathological changes and neurotoxicity

allowing for the transport of Aβ aggregates (and possibly other seeds) from the periphery to the brain raises the speculation that environmental amyloidogenic seeds might act as risk factors for certain neurodegenerative diseases.

Historical Perspective

For over 40 years, certain similarities between prion diseases and Alzheimer´s disease (AD) have evoked speculation that AD might be infectious in a manner similar to the prionoses (Brown et al. 1982; Prusiner 1984). To test this hypothesis, nonhuman primates were inoculated intracerebrally with brain material from AD patients, but the results were not conclusive (Goudsmit et al. 1980). The intracerebral inoculation of buffy coat from the blood of AD patients into hamsters was reported to induce prion-like spongiform encephalopathy (Manuelidis et al. 1988), but follow-up work (Godec et al. 1991) failed to replicate this finding. In contrast, subsequent papers have shown that the intracerebral inoculation of crude homogenates from AD brain into marmosets can induce β-amyloid deposition after incubation times of at least 5–6 years (Baker et al. 1993; Ridley et al. 2006).

In vitro, the aggregation of the prion protein (PrP) and the amyloid-β peptide (Aβ) is a nucleation-dependent polymerization process (Fig. 1; Harper and Lansbury 1997), a phenomenon that is potentially applicable to all amyloids (Eisenberg and Jucker 2012). Specifically, in the disease state, the conformation of a particular protein is driven toward an abnormal three-dimensional architecture by a process of corruptive protein templating. Thus, in vitro studies predict that the experimental induction of Aβ aggregation and deposition should also be possible in vivo, given that the seed, host and incubation time are well chosen (Walker et al. 2006a). The development of genetically defined and specific pathogen-free mouse models of β-amyloidosis has provided a more efficient and definitive mean of testing the hypothesis that AD-like lesions can be seeded in vivo. Studies in our laboratories and then by others have shown that the deposition of Aβ can be instigated in the brains of Aβ-precursor protein (APP)-transgenic mice by the intracerebral infusion of brain extracts containing minute amounts of aggregated (multimeric) Aβ (Kane et al. 2000; Meyer-Luehmann et al. 2006; Eisele et al. 2009; Langer et al. 2011; Morales et al. 2012; Watts et al. 2011; Hamaguchi et al. 2012).

Nature of the β-Amyloid-Inducing Agent

Burgeoning evidence suggests that aggregated Aβ in the brain extract is essential for its β-amyloid-inducing activity and that the β-amyloid-induction occurs via an Aβ-templating mechanism. The arguments are as follows: (1) Aβ-induction is time-dependent; the longer the incubation time, the greater the burden of induced β-amyloid deposits (Fig. 2; Meyer-Luehmann et al. 2006; Hamaguchi et al. 2012); (2) Aβ-induction is dependent on the concentration of Aβ in the extract (Meyer-Luehmann et al. 2006); (3) immunodepletion of Aβ from the brain extract hinders seeding (Meyer-Luehmann et al. 2006); (4) seeding can be blocked by denaturation of the extract with formic acid (Meyer-Luehmann et al. 2006); (5) Aβ deposition cannot be seeded in non-transgenic host mice because murine Aβ differs from that in humans at three amino acid positions and therefore does not aggregate as readily as does human-sequence Aβ (Kane et al. 2000; Meyer-Luehmann et al. 2006); (6) the induced Aβ deposits morphologically reflect the deposits in the donor extract (Meyer-Luehmann et al. 2006); and (7) Aβ-rich brain extracts from transgenic mice seed as potently as do AD brain extracts (Meyer-Luehmann et al. 2006), rendering it unlikely that factors such as a cross-species immune reaction or human-specific microbes or viruses are involved.

Although higher molecular-weight aggregates of Aβ can seed effectively, recent evidence indicates that the inductive Aβ seeds are diverse and can occur both as proteinase K (PK)-resistant aggregates in the insoluble fraction and as soluble, PK-sensitive small multimers in the $100,000 \times g$ soluble fraction (Langer et al. 2011). Extended sonication, and thus presumed fragmentation of the extract and generation of more seeds (Fig. 1), further indicates that the Aβ aggregates comprise a continuum of β-amyloid-inducing agents of various sizes (Langer et al. 2011).

In addition, purified Aβ fibrils from brain homogenates as well as aggregated synthetic Aβ have been shown to induce cerebral Aβ deposition in APP transgenic mice (Stöhr et al. 2012). However, synthetic Aβ fibrils needed higher concentration for the induction and were therefore far less potent compared to naturally generated Aβ aggregates (Meyer-Luehmann et al. 2006; Stöhr et al. 2012). Similarly, transmitting prion diseases with in vitro generated PrP is also relatively inefficient (Legname et al. 2004; Wang et al. 2010). These observations support growing evidence that Aβ, like prions, can misfold into polymorphic and polyfunctional strains (Petkovà et al. 2005; Eisenberg and Jucker 2012) although an unknown brain-derived co-factor that potentiates the seeding activity of brain-derived material cannot be excluded.

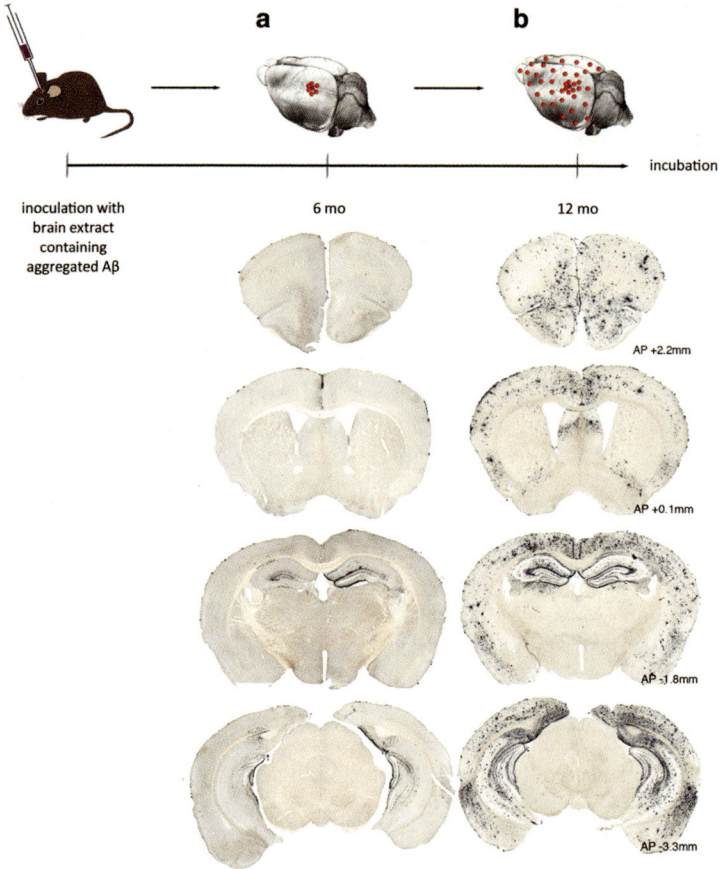

Fig. 2 Induction and spread of Aβ lesions in an APP transgenic mouse model after intracerebral application of β-amyloid-containing brain extract. (**a**) The intracerebral inoculation of brain extract containing aggregated Aβ into the hippocampus and overlying neocortex induces cerebral β-amyloidosis. The amyloid induction is time-dependent. After a 6-month incubation period, the R1.40 mouse model (Hamaguchi et al. 2012) manifests Aβ deposition confined to the hippocampus and the injected neocortical area. (**b**) A longer incubation time (12 months) results in a remarkable spreading of Aβ deposition to neighboring brain regions and throughout the entire forebrain

Inductive Spreading of β-Amyloid Aggregates

Cross-sectional analyses of postmortem human brains reveal a characteristic progression of β-amyloid plaques from the neocortex to subcortical regions (Thal et al. 2006). Moreover, the progression of cerebral β-amyloidosis often corresponds to functionally and anatomically coupled brain regions (Buckner et al. 2005; Walker and Jucker 2011), suggesting that Aβ aggregation may spread by templated Aβ misfolding and seeding.

Indeed, following local intracerebral application of Aβ-containing extract, Aβ deposits appear first at the injection site and subsequently appear to proliferate throughout the injected brain region (Fig. 2; Eisele et al. 2009; Hamaguchi et al. 2012). Moreover, after infusion of the Aβ-containing extract into the hippocampus, Aβ deposits are selectively induced in the entorhinal cortex and vice versa, suggesting that Aβ deposits can be induced along defined neuronal pathways (Eisele et al. 2009; Walker et al. 2002; Jucker and Walker 2011). In addition, the infusion of donor extract into one brain region can precipitate Aβ deposits in the vasculature of anatomically separate locations, suggesting that the seeds might also migrate along perivascular fluid drainage channels and/or within the bloodstream (Walker and Jucker 2011).

The transport of Aβ seeds is unequivocally demonstrated by the finding that Aβ deposits in the brains of APP mice can be induced by the peripheral infusion of seeds into the peritoneal cavity (Eisele et al. 2010). More recently, cerebral β-amyloidosis has been reported to be induced in transgenic mice by blood trans-fusion experiments (C. Soto, personal communication), suggesting intravascular transport of Aβ seeds.

Similar Principles for Tau Lesions?

Neurofibrillary tangles appear and ramify in an even more stereotypical pattern than do Aβ lesions. Neurofibrillary tangles first arise in the locus coeruleus and entorhinal/ limbic brain areas and then spread to interconnected neocortical regions (Braak and Braak 1991; Braak and Del Tredici 2011), presumably by neuronal transport and trans-synaptic spread. Similar to the inoculation protocol using Aβ-containing brain extracts, neurofibrillary (tau) tangles have been induced by the intracerebral infusion of brain extracts containing abnormal tau filaments into mice bearing human tau transgenes, even though the mice do not normally develop tau lesions (Clavaguera et al. 2009). The induction of tauopathy is again time- and brain region-dependent, and immunodepletion of tau from the donor brain extract prevents seeding (Clavaguera et al. 2009). Spreading of the induced tau lesions occurs from the injection site to more remote, anatomically connected brain regions (Clavaguera et al. 2009), a conclusion corroborated by recent studies of the progression of tau lesions in mice with region-specific tau expression (Liu et al. 2012; Calignon et al. 2012). Several in vitro studies also support a prion-like propagation of tau lesions (Frost et al. 2009a; Goedert et al. 2010). Moreover, there is in vitro evidence of polymorphic tau lesions that propagate by templated conversion (e.g., Frost et al. 2009b; Guo and Lee 2011).

Conclusions

1. Cerebral β-amyloidosis and neurofibrillary tangles can be induced in transgenic mice by a single inoculation of a dilute β-amyloid- and neurofibrillary tangle-containing brain extract, respectively.
2. Induced β-amyloid and tau lesions spread within and among brain regions.
3. Specific conformations of the respective protein aggregates (possibly with the participation of unknown co-factors) act as the β-amyloid-inducing or tau-inducing agents ('seeds').
4. There is evidence that conformational variants of Aβ and tau can be propagated *in vivo* (Aβ 'strains', tau 'strains').
5. Small and soluble proteopathic seeds appear to be highly effective.

Implications and Speculations

While prion disease is the only demonstrably infectious neurodegenerative disease, accumulating evidence indicates that a variety of neurological and systemic amyloid disorders are instigated and propagated by molecular mechanisms resembling those that characterize the prionoses (Soto et al. 2006; Walker et al. 2006a, b; Frost and Diamond 2010; Goedert et al. 2010; Brundin et al. 2010; Jucker and Walker 2011). Apart from Aβ and tau, convincing in vivo evidence also has been published for the prion-like induction of α-synucleinopathy (Kordower et al. 2008; Li et al. 2008; Hansen et al. 2011; Mougenot et al. 2012; Luk et al. 2012a, b). Thus, this emerging principle of pathogenesis has the potential to unify experimental and therapeutic approaches to seemingly disparate disorders.

In AD, β-amyloid deposition begins in the brain at least a decade before the onset of cognitive decline and hence is an early indicator of the disease (Holtzman et al. 2011). For this reason, it is probable that an effective disease-modifying therapy must be initiated long before the disease has caused permanent damage to the brain (Selkoe 2011; Golde et al. 2011). Soluble Aβ seeds may arise even sooner than do the classical Aβ deposits in senile plaques and in the walls of cerebral blood vessels. The finding that variants of Aβ seeds govern the type (and possibly toxicity) of Aβ-aggregates could explain the heterogeneous Aβ morphotypes in brain (Levine and Walker 2010; Eisenberg and Jucker 2012), which in turn may underlie the heterogeneous pathology and progression of AD. Thus, Aβ seeds and factors that influence the conformational characteristics of these pathogenic agents represent promising therapeutic targets.

The infusion of Aβ-rich brain extracts into the peritoneal cavity of young APP transgenic mice induces Aβ deposition in the brain (Eisele et al. 2010), indicating that—at least in transgenic mice—Aβ seeds, like prions, can reach the central nervous system from the periphery. Because there is still no direct evidence that non-prion neurodegenerative diseases can be caused by seeds arising from systemic sources or from the environment (e.g., in food or air), the practical implications of this finding are uncertain. However, a better understanding of the trafficking of pathogenic seeds is likely to yield new insights into the endogenous evolution of the disease. Are there heterologous amyloidogenic seeds in our environment that can be taken up and transported to the brain, where they trigger cerebral β-amyloidosis by cross-seeding (Eisenberg and Jucker 2012)? Experimentally, 'cross-seeding' is generally less potent than is homologous seeding, but the potential corruption of proteins by exogenous material with amyloid-like structural properties appears possible.

Further studies are required to establish the clinical and epidemiological implications of this prion-like aspect of AD pathology. The available information indicates that it is unlikely that AD and other cerebral amyloidoses are communicable under ordinary circumstances. However, it is conceivable that the emergence of these disorders might be facilitated under certain extraordinary conditions. Both prion disease and Aβ deposition are most efficiently seeded by direct introduction of the seeding agent into the brain. β-Amyloid can be induced in young APP transgenic mice by intracerebrally implanted stainless steel wires that have been coated with brain extract containing aggregated Aβ (Eisele et al. 2009). While this study demonstrates the theoretical possibility that cerebral β-amyloidosis might be actuated in humans by surgical instruments, there is currently no evidence that this is the case. However, proof of such a phenomenon could be obscured by the very long incubation period that characterizes neurodegenerative diseases such as AD. Clearly, more data are needed, including careful epidemiological studies, to provide definitive answers to these questions.

References

Baker HF, Ridley RM, Duchen LW, Crow TJ, Bruton CJ (1993) Evidence for the experimental transmission of cerebral beta-amyloidosis to primates. Int J Exp Pathol 74:441–454

Braak H, Braak E (1991) Neuropathological stageing of Alzheimer-related changes. Acta Neuropathol 82:239–259

Braak H, Del Tredici K (2011) The pathological process underlying Alzheimer's disease in individuals under thirty. Acta Neuropathol 121:171–181

Brown P, Salazar AM, Gibbs CJ, Gajdusek DC (1982) Alzheimer's disease and transmissible virus dementia (Creutzfeldt-Jakob disease). Ann NY Acad Sci 396:131–143

Brundin P, Melki R, Kopito R (2010) Prion-like transmission of protein aggregates in neurodegenerative diseases. Nat Rev Mol Cell Biol 11:301–307

Buckner RL, Snyder AZ, Shannon BJ, LaRossa G, Sachs R, Fotenos AF, Sheline YI, Klunk WE, Mathis CA, Morris JC, Mintun MA (2005) Molecular, structural, and functional characterization of Alzheimer's disease: evidence for a relationship between default activity, amyloid, and memory. J Neurosci 225:7709–7717

Calignon A, Polydoro M, Suarez-Calvet M, William C, Adamowicz DH, Kopeikina KJ, Pitstick R, Sahara N, Ashe KH, Calson GA, Spires-Jones TL, Hyman BT (2012) Propagation of Tau pathology in a model of early Alzheimer's disease. Neuron 73:685–697

Clavaguera F, Bolmont T, Crowther RA, Abramowski D, Frank S, Probst A, Fraser G, Stalder AK, Beibel M, Staufenbiel M, Jucker M, Goedert M, Tolnay M (2009) Transmission and spreading of tauopathy in transgenic mouse brain. Nat Cell Biol 11:909–913

Eisele YS, Bolmont T, Heikenwalder M, Langer F, Jacobson LH, Yan ZX, Roth K, Aguzzi A, Staufenbiel M, Walker LC, Jucker M (2009) Induction of cerebral beta-amyloidosis: intracerebral versus systemic Abeta inoculation. Proc Natl Acad Sci USA 106:12926–12931

Eisele YS, Obermüller U, Heilbronner G, Baumann F, Kaeser SA, Wolburg H, Walker LC, Staufenbiel M, Heikenwalder M, Jucker M (2010) Peripherally applied Abeta-containing inoculates induce cerebral beta-amyloidosis. Science 330:980–982

Eisenberg D, Jucker M (2012) The amyloid state of proteins in human diseases. Cell 148: 1188–1203

Frost B, Diamond MI (2010) Prion-like mechanisms in neurodegenerative diseases. Nat Rev Neurosci 11:155–159

Frost B, Jacks RL, Diamond MI (2009a) Propagation of tau misfolding from the outside to the inside of a cell. J Biol Chem 284:12845–12852

Frost B, Ollesch J, Wille H, Diamond MI (2009b) Conformational diversity of wild-type Tau fibrils specified by templated conformation change. J Biol Chem 284:3546–3551

Godec MS, Asher DM, Masters CL, Kozachuk WE, Friedland RP, Gibbs CJ Jr, Gajdusek DC, Rapoport SI, Schapiro MB (1991) Evidence against the transmissibility of Alzheimer's disease. Neurology 41:1320

Goedert M, Clavaguera F, Tolnay M (2010) The propagation of prion-like protein inclusions in neurodegenerative diseases. Trends Neurosci 33:317–325

Golde TE, Schneider LS, Koo EH (2011) Anti-aβ therapeutics in Alzheimer's disease: the need for a paradigm shift. Neuron 69:203–213

Goudsmit J, Morrow CH, Asher DM, Yanagihara RT, Masters CL, Gibbs CJ Jr, Gajdusek DC (1980) Evidence for and against the transmissibility of Alzheimer disease. Neurology 30: 945–950

Guo JL, Lee VM-Y (2011) Seeding of normal Tau by pathological Tau conformers drives pathogenesis of Alzheimer-like tangles. J Biol Chem 286:15317–15331

Hamaguchi T, Eisele YS, Varvel NH, Lamb BT, Walker LC, Jucker M (2012) The presence of Aβ seeds, and not age per se, is critical to the initiation of Aβ deposition in the brain. Acta Neuropathol 123:31–37

Hansen C, Angot E, Bergström AL, Steiner JA, Pieri L, Paul G, Outeiro TF, Melki R, Kallunki P, Fog K, Li JY, Brundin P (2011) α-Synuclein propagates from mouse brain to grafted dopaminergic neurons and seeds aggregation in cultured human cells. J Clin Invest 18:1–11

Harper JD, Lansbury PT (1997) Models of amyloid seeding in Alzheimer's disease and scrapie: mechanistic truths and physiological consequences of the time-dependent solubility of amyloid proteins. Annu Rev Biochem 66:385–407

Holtzman DM, Morris JC, Goate AM (2011) Alzheimer's disease: the challenge of the second century. Sci Transl Med 3:77sr1, Review

Jucker M, Walker LC (2011) Pathogenic protein seeding in Alzheimer disease and other neurodegenerative disorders. Ann Neurol 70:532–540

Kane MD, Lipinski WJ, Callahan MJ, Bian F, Durham RA, Schwarz RD, Roher AE, Walker LC (2000) Evidence for seeding of beta-amyloid by intracerebral infusion of Alzheimer brain extracts in beta-amyloid precursor protein-transgenic mice. J Neurosci 20:3606–3611

Kordower JH, Chu Y, Hauser RA, Freeman TB, Olanow CW (2008) Lewy body-like pathology in long-term embryonic nigral transplants in Parkinson's disease. Nat Med 14:504–506

Langer F, Eisele YS, Fritschi SK, Staufenbiel M, Walker LC, Jucker M (2011) Soluble A{beta} seeds are potent inducers of cerebral {beta}-amyloid deposition. J Neurosci 31:14488–14495

Legname G, Baskakov IV, Nguyen HO, Riesner D, Cohen FE, DeArmond SJ, Prusiner SB (2004) Synthetic mammalian prions. Science 305:673–676

Levine H 3rd, Walker LC (2010) Molecular polymorphism of Abeta in Alzheimer's disease. Neurobiol Aging 31:542–548

Li JY, Englund E, Holton JL, Soulet D, Hagell P, Lees AJ, Lashley T, Quinn NP, Rehncrona S, Björklund A, Widner H, Revesz T, Lindvall O, Brundin P (2008) Lewy bodies in grafted neurons in subjects with Parkinson's disease suggest host-to-graft disease propagation. Nat Med 14:501–503

Liu L, Drouet V, Wu JW, Witter MP, Small SA, Clelland C, Duff K (2012) Trans-synaptic spread of tau pathology in vivo. PLoS One 7:e31302

Luk KC, Kehm VM, Zhang B, O'Brien P, Trojanowski JQ, Lee VM-Y (2012a) Intracerebral inoculation of pathological α-synuclein initiates a rapidly progressive neurodegenerative α-synucleinopathy in mice. J Exp Med 209(5):975–986

Luk KC, Kehm V, Carroll J, Zhang B, O'Brien P, Trojanowski JQ, Lee VM (2012b) Pathological α-synuclein transmission initiates Parkinson-like neurodegeneration in nontransgenic mice. Science 338(6109):949–953

Manuelidis EE, de Figueiredo JM, Kim JH, Fritch WW, Manuelidis L (1988) Transmission studies from blood of Alzheimer disease patients and healthy relatives. Proc Natl Acad Sci USA 85: 4898–4901

Meyer-Luehmann M, Coomaraswamy J, Bolmont T, Kaeser S, Schaefer C, Kilger E, Neuenschwander A, Abramowski D, Frey P, Jaton AL, Vigouret JM, Paganetti P, Walsh DM, Mathews PM, Ghiso J, Staufenbiel M, Walker LC, Jucker M (2006) Exogenous induction of cerebral beta-amyloidogenesis is governed by agent and host. Science 313:1781–1784

Morales R, Duran-Aniotz C, Castilla J, Estrada LD, Soto C (2012) De novo induction of amyloid-β deposition in vivo. Mol Psychiatry 17:1347–1353

Mougenot AL, Nicot S, Bencsik A, Morignat E, Verchere J, Lakhdar L, Legastelois S, Baron T (2012) Prion-like acceleration of a synucleinopathy in a transgenic mouse model. Neurobiol Aging 33:2225–2228

Petkova AT, Leapman RD, Guo Z, Yau WM, Mattson MP, Tycko R (2005) Self-propagating, molecular-level polymorphism in Alzheimer's beta-amyloid fibrils. Science 307:262–265

Prusiner SB (1984) Some speculations about prions, amyloid, and Alzheimer's disease. N Engl J Med 310:661–663

Ridley RM, Baker HF, Windle CP, Cummings RM (2006) Very long term studies of the seeding of beta-amyloidosis in primates. J Neural Transm 113:1243–1251

Selkoe DJ (2011) Resolving controversies on the path to Alzheimer's therapeutics. Nat Med 17: 1060–1065

Soto C, Estrada L, Castilla J (2006) Amyloids, prions and the inherent infectious nature of misfolded protein aggregates. Trends Biochem Sci 31:150–155

Stöhr J, Watts JC, Mensinger ZL, Oehler A, Grillo SK, DeArmond SJ, Prusiner SB, Giles K (2012) Purified and synthetic Alzheimer's amyloid beta (Aβ) prions. Proc Natl Acad Sci USA 109(27):11025–11030

Thal DR, Capetillo-Zarate E, Del Tredici K, Braak H (2006) The development of amyloid beta protein deposits in the aged brain. Sci Aging Knowledge Environ 2006:re1

Walker LC, Jucker M (2011) Amyloid by default. Nat Neurosci 14:669–670

Walker LC, Callahan MJ, Bian F, Durham RA, Roher AE, Lipinski WJ (2002) Exogenous induction of cerebral beta-amyloidosis in betaAPP-transgenic mice. Peptides 23:1241–1247

Walker LC, Levine H, Mattson MP, Jucker M (2006a) Inducible proteopathies. Trends Neurosci 29:438–443

Walker L, Levine H, Jucker M (2006b) Koch's postulate and infectious proteins. Acta Neuropathol 112:1–4

Wang F, Wang X, Yuan C-G, Ma J (2010) Generating a prion with bacterially expressed recombinant prion protein. Science 327:1132–1135

Watts JC, Giles K, Grillo SK, Lemus A, DeArmond SJ, Prusiner SB (2011) Bioluminescence imaging of Abeta deposition in bigenic mouse models of Alzheimer's disease. Proc Natl Acad Sci USA 108:2528–2533

Amyloid-β Transmissibility

C. Duran-Aniotz, R. Morales, I. Moreno-Gonzalez, and C. Soto

Abstract Alzheimer's disease (AD) is the most common type of dementia in the elderly population. This disorder is histopathologically characterized by the presence of cerebral deposits of fibrillar aggregates consisting of amyloid-β protein (Aβ). Aβ aggregates have been characterized in detail, placing them as the main factors responsible for the deleterious clinical features observed in this disease. Interestingly, protein misfolding and aggregation are the predominant pathological events in several other diseases known as protein misfolding disorders (PMDs). PMDs include AD, Parkinson's disease, Huntington's disease, type-2 diabetes and Transmissible Spongiform Encephalopathies, or prion diseases, among others. Prion diseases are a group of disorders that can be transmitted by a proteinaceous infectious material termed prion. Compelling studies have demonstrated the infectious properties of misfolded prion aggregates that replicate following a seeding-nucleation mechanism. Recent experiments performed in animal models of diverse PMDs have shown that misfolded aggregates can induce the disease process by a prion-like mechanism, revealing a potential infectious origin for some of these diseases. In this chapter, we will review the recent studies showing that AD-like pathology can be induced in a similar way to prion diseases.

C. Duran-Aniotz
Mitchell Center for Alzheimer's Disease and Related Brain Disorders, Department of Neurology, University of Texas Houston Medical School, 6431 Fannin, MSE Room 422, Houston, TX 77030, USA

Universidad de los Andes, Facultad de Medicina, Av. San Carlos de Apoquindo 2200, Las Condes, Santiago, Chile

R. Morales • I. Moreno-Gonzalez • C. Soto (✉)
Mitchell Center for Alzheimer's Disease and Related Brain Disorders, Department of Neurology, University of Texas Houston Medical School, 6431 Fannin, MSE Room 422, Houston, TX 77030, USA
e-mail: claudio.soto@uth.tmc.edu

M. Jucker and Y. Christen (eds.), *Proteopathic Seeds and Neurodegenerative Diseases*,   71
Research and Perspectives in Alzheimer's Disease,
DOI 10.1007/978-3-642-35491-5_6, © Springer-Verlag Berlin Heidelberg 2013

Introduction: Protein Misfolding Disorders

Proteins need to acquire their native folding to perform a correct biological function. Abnormal folding of proteins induces unusual biological properties (Koo et al. 1999; Soto 2003; Thomas et al. 1995). Misfolded proteins can be produced by cellular stress, mutations, errors in protein synthesis and other cellular alterations, leading either to a loss of normal function or to toxic properties, both triggering cell malfunction, cell death, tissue damage and disease (Luheshi et al. 2008; Soto 2003). Protein misfolding disorders (PMDs) include several brain and systemic amyloidoses such as Alzheimer's disease (AD), Parkinson's disease (PD), Huntington's disease (HD), amyotrophic lateral sclerosis (ALS), type-2-diabetes, systemic amyloidosis and Transmissible Spongiform Encephalopathies (TSEs), or Prion diseases (Luheshi et al. 2008; Soto 2003).

The presence of misfolded protein deposits in affected tissues is a general characteristic of all PMDs (Luheshi et al. 2008; Soto 2003). Current evidence suggests that the deposition of misfolded protein aggregates is a critical event triggering tissue damage and pathogenesis in these diseases (Luheshi et al. 2008; Soto 2003). Despite the fact that each PMD is characterized by the deposition of aggregates formed by different proteins [amyloid-beta ($A\beta$) and tau proteins in AD, misfolded prion protein (PrP^{Sc}) in TSEs, α-synuclein (α-syn) in PD, huntingtin in HD, etc.)], comparable morphologic, biologic and biochemical features have been described for all of them (Soto 2003). The typical molecular arrangement of these amyloid aggregates consists of clusters of misfolded proteins organized in cross-β-sheet structures (Blake et al. 1996). This particular structure provides several properties, including toxicity and resistance to clearance mechanisms.

Although the majority of PMDs cases are sporadic in nature, inherited diseases provide significant insight about the genes contributing to these disorders. Strikingly, in all PMDs, mutations in the gene encoding for the proteins that misfold and aggregate are associated with familial forms of the disease (Hardy and Gwinn-Hardy 1998). In vitro aggregation profiles of mutant proteins associated with familial PMDs are more aggressive compared to their "natural" counterparts. Furthermore, the generation of transgenic animals expressing proteins harboring point mutations associated with inherited PMDs generally results in neuropathological and clinical features similar to the ones found in human disorders (Aguzzi et al. 1994; Chapman et al. 2001; McGowan et al. 2006; Rockenstein et al. 2007). In the case of AD, several models have been developed involving proteins that directly ($A\beta$, tau; Allen et al. 2002; Games et al. 1995; Hsiao et al. 1996) or indirectly (α-syn, presenilins, ApoE; Brendza et al. 2002; Duff et al. 1996) participate in neurodegeneration. Various transgenic mice expressing the human amyloid precursor protein (APP) have been developed, mimicking the brain $A\beta$ aggregation observed in AD patients. Among them, APP23 over-expresses the human APP_{751} harboring the Swedish mutation (Sturchler-Pierrat et al. 1997). As a result, these mice exhibit brain $A\beta$ deposits starting at around 6 months of age and extensive amyloid aggregation in neocortex and hippocampus at 24 months (Calhoun et al. 1998, 1999; Sturchler-Pierrat et al. 1997). Perhaps the most used and best-characterized AD mouse model is the one known as tg2576 (Hsiao et al. 1996). This mouse strain, also producing human APP harboring the Swedish mutation, starts Aβ deposition and memory impairments between 8 and 9 months (Kawarabayashi et al. 2001). Among several other transgenic models of AD, mice carrying multiple mutations in different AD-associated proteins have been generated, mimicking an aggravated neuropathological scenario. For example, $APP_{Swe}/PS1\Delta E9$ mice expressing mutated human APP and presenilin-1 (PS1) (Borchelt et al. 1997) show the occasional presence of Aβ aggregates as early as 5–6 months and widespread amyloid plaque accumulation by 12 months. The presence of Aβ aggregates and the number of AD-related mutations have revealed a positive correlation (Clinton et al. 2010; Oakley et al. 2006; Oddo et al. 2003), demonstrating the fundamental contribution of misfolded proteins to the pathogenesis of AD.

Misfolded Aβ Aggregates and AD

AD is the most important PMD due to its prevalence in the aged population. This disease, described for the first time by Alois Alzheimer in 1907, is characterized by a progressive and irreversible brain degeneration that leads to gradual memory loss and decline of cognition, including changes in personality. These abnormalities are induced by synaptic dysfunction and neuronal loss. AD is the most common type of dementia in the elderly, accounting for 60–80 % of all dementia cases, which also include frontotemporal dementia, dementia with Lewy bodies and vascular dementia, among others (Ballard et al. 2011). Currently, more than 26 million people worldwide suffer from AD (Brookmeyer et al. 2007) and about 10 % of individuals over the age of 65 have some degree of AD; the frequency increases to nearly 50 % by the age of 85. Since the proportion of elderly people is currently increasing, and the risk of developing AD escalates exponentially with age, it is predicted that the prevalence of this dementia will double by 2050 (Brayne 2007). At this moment, there are no effective methods to treat or diagnose AD, which constitutes a major problem for public health.

As we have previously noted, the primary pathological hallmark in the AD brain is the presence of amyloid plaques and neurofibrillary tangles (NFT), which are composed of extracellular deposition of misfolded Aβ peptides and intracellular hyperphosphorylated tau, respectively (Glenner and Wong 1984; Grundke-Iqbal et al. 1986). Senile plaques are amyloidogenic deposits with a dense central core, surrounded by dystrophic neurites and a severe inflammatory reaction. NFTs are abnormal, filamentous inclusions found in neuron somata and composed principally of abnormally folded tau protein. Tau, a microtubule-associated protein, forms paired helical filaments upon hyperphosphorylation, which leads to impaired axonal transport and, eventually, cell death (Brandt et al. 2005; Vossel et al. 2010). Other AD lesions include loss of synaptic connection, neuronal death, cerebral amyloid angiopathy and oxidative stress (Serrano-Pozo et al. 2011). Inflammation in the AD brain is characterized by activated microglia and reactive astrocytes that are associated with Aβ deposits and up-regulation of many mediators of the inflammatory response (McGeer and McGeer 2007). The medial temporal lobe areas, e.g., hippocampus and entorhinal cortex, are the first regions affected by the pathological features of AD (Braak and Braak 1991).

A widely accepted model explaining brain degeneration leading to the clinical signs of AD is the amyloid cascade hypothesis. This model was proposed 20 years ago, shortly after the initial genetic discoveries of APP mutations leading to familial AD (Hardy 1992). The model posits that abnormal aggregation of Aβ and its deposition in the brain parenchyma are the decisive steps that trigger AD pathology. NFTs, cell loss, inflammation, and dementia are the result of Aβ pathology. However, several reports suggest that there is not a positive correlation between extracellular fibrillar amyloid burden and cognitive impairment or dementia (Arriagada et al. 1992; Price et al. 2009). To accommodate for recent findings, the amyloid hypothesis has been modified to propose that non-fibrillar, soluble Aβ oligomers, rather than insoluble Aβ present in plaques, are responsible for initiating AD pathology (Glabe 2006; Haass and Selkoe 2007). In support of this model, levels of soluble Aβ correlate better with the degree of dementia in humans than amyloid deposits (Lesne et al. 2006). Current information suggests that soluble oligomers could be directly responsible for the primary neurotoxic effects and memory impairments, particularly during pre-symptomatic stages of the disease (Haass and Selkoe 2007). Small aggregates composed of $A\beta_{40}$ and $A\beta_{42}$ fragments have been reported as cytotoxic in vivo an in vitro (Cleary et al. 2004; Reed et al. 2011; Shankar et al. 2007; Townsend et al. 2006; Walsh et al. 2002). Exposure to soluble and aggregated Aβ in neuronal and organotypic cultures induces cellular dysfunction and cell death. Animals injected with various synthetic or brain-derived $A\beta_{42}$ oligomers develop neuronal damage, alterations in long-term potentiation and behavior impairments. The identification of oligomers as harmful molecules has opened new research pathways for developing innovative therapeutic strategies to combat AD (Klein et al. 2001; Walsh and Selkoe 2007).

Although the molecular bases of AD have been extensively studied, the events triggering the pathology still remain unknown in most of the cases. AD is a heterogeneous and multifactorial disease in which the underlying etiology is not yet clear. A small proportion of AD cases, so-called familial AD, shows hereditary transmission of the disease, affecting genes encoding APP, PS1 and presenilin 2 (PS2; Goate et al. 1991; Levy-Lahad et al. 1995; Schellenberg et al. 1992). Mutations in each of these genes result in elevated levels of Aβ, which is deposited in the brains of AD patients as senile plaques. However, sporadic cases of AD represent around 95 % of all cases of AD. The specific events triggering sporadic disease are unknown, but there are several risk factors that can increase the probability of developing AD (Reitz et al. 2011). The main risk factor for AD is aging. The presence of some other risk factors described for AD [ApoE polymorphisms (Roses and Saunders 1994), co-existence of other diseases (Morales et al. 2009)] is neither required nor enough to trigger AD. Recent evidence in animal models suggesting prion-like spreading of protein misfolding in AD (see below) could explain a proportion of sporadic cases of AD with an unknown etiology.

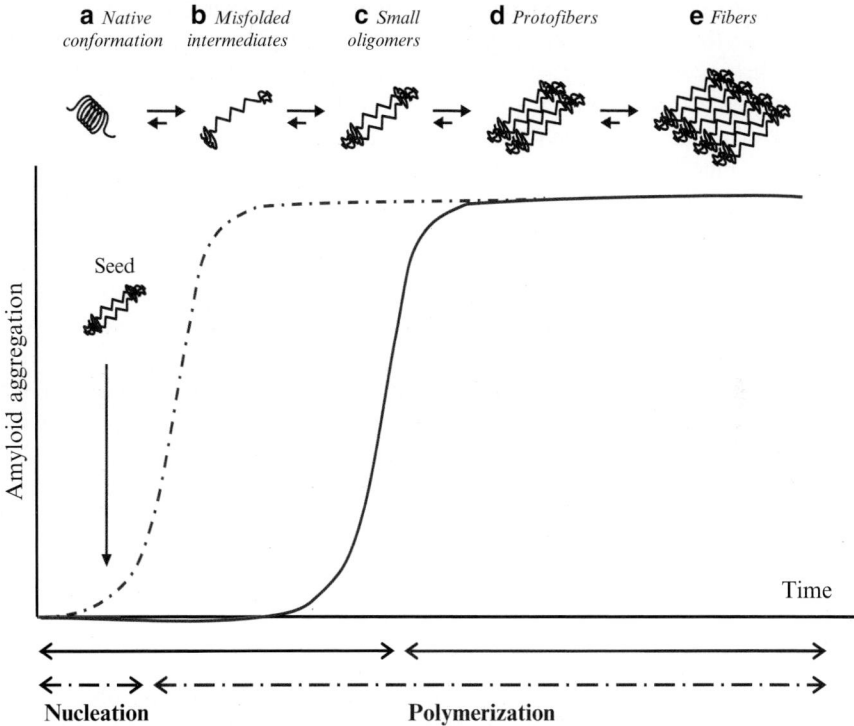

Fig. 1 Nucleation-polymerization model. Native folding of proteins is needed to perform a correct biological function (A). Various types of alterations, including changes in the cellular environment, somatic mutations, and errors in protein synthesis, may result in partially denatured states. (B) Misfolded proteins are usually prone to aggregate, forming small oligomers (C), which can keep growing by integration of new protein molecules in the ends to form larger structures termed protofibrils (D) and mature fibrils (E). The process of aggregation follows a seeding/nucleation kinetic model (represented by a *solid line*), which consists of two distinct phases. The first stage, or "lag phase", involves the formation of the first misfolded units or "seeds". In this phase, the growth is slow and corresponds to the rate-limiting step for further aggregation. Subsequently, in the second phase, or "elongation phase", the rapid recruitment of new soluble units into the growing protein polymer takes place. The aggregation process can be accelerated by the addition of previously formed seeds (*dashed line*)

Mechanisms and Intermediates in Protein Misfolding Aggregation: The "Seeding" Model

Amyloid formation is initiated by the conversion of native (probably partially denatured) soluble proteins into insoluble large aggregates consisting of cross-β-sheet structures. Amyloid formation follows a classical kinetic pathway, often referred to as the seeding/nucleation model of polymerization (Jarrett and Lansbury 1993; Soto et al. 2006). This process involves two different phases, as described in Fig. 1. The first stage, or "lag phase," involves the slow formation of the initial, stable misfolded aggregates. Small oligomeric structures (referred to as "seeds") are formed in this phase through unfavorable intermolecular interactions of monomers. In the lag phase, the growth is slow and corresponds to the rate-limiting step leading to further aggregation. Subsequently, in the second stage, or "elongation phase," the recruitment of new units into the growing aggregates takes place in a much-accelerated fashion until a plateau level of polymerization is attained. This "seeding-nucleation" process can be accelerated by the addition of previously formed seeds that provide a nucleus for the incorporation of soluble components, promoting the rapid fibrillar growth (Jarrett and Lansbury 1993; Soto et al. 2006). The resulting misfolded protein aggregates cover a wide range of sizes, including soluble small oligomeric units, protofibrils and fibrils (large aggregates).

In vitro studies using diverse proteins implicated in PMDs have shown that the addition of a small amount of pre-formed aggregates decreases the extent of the lag phase in a concentration-dependent manner (Jarrett and Lansbury 1993; Soto et al. 2006). Interestingly, oligomers seem to be the best seeds to promote faster aggregate formation (Cheon et al. 2008). Even though the proteins associated with each PMD are different, the great similarities in the mechanisms of polymerization suggest common pathways associated with the induction of the pathological features.

Proteins Behaving as Pathogens: The Case of Prion Diseases

Prions are the proteinaceous infectious agents responsible for TSEs (Aguzzi and Calella 2009; Prusiner 1998). Although uncommon in humans, prion diseases affect several mammalian species. Infamous are the cases of prion diseases in cattle (with the potential to transmit the disease to humans), sheep (causing serious damage to the food industry), or cervids (affecting a wide range of captive and wild animals; Collinge 2001; Sigurdson and Aguzzi 2006). The short clinical phase preceded by a long and silent asymptomatic stage in which the prion agent is replicating in the body are hallmark features of prion diseases. The lack of diagnostic methods and effective therapeutic strategies makes these diseases a serious public health threat (Soto 2004).

Prion diseases were first described in animals as "long incubation-infectious diseases able to produce neurological disorders". Unlike the case of AD, prion diseases are characterized by a short clinical stage that is invariably fatal. Prions are able to transmit disease by an unorthodox mechanism that defies the central dogma of molecular biology: the information required to transmit disease is encoded by a misfolded form of a normally produced protein, PrP^{Sc}, which is proficient in transmitting its conformation to the normally produced protein, termed PrP^{C} (Prusiner 1998). The accumulation of PrP^{Sc} aggregates in the brain is thought to be the main factor responsible for brain degeneration, cell death and the clinical signs characteristics of this group of diseases (Aguzzi and Calella 2009). The molecular bases of prion diseases are very similar to those observed for other amyloidoses. The previously described seeding-nucleation model neatly accounts for the mechanism of prion replication, with misfolded oligomeric seeds being the infectious agent that propagate the disease by inducing the conversion of the normal prion protein into more of the toxic agent (Soto et al. 2006). The seeding-nucleation model for prion infectivity provides feasible explanations for some of the unique features of prions, such as the fidelity of prion strain propagation and the molecular mechanism of the species barrier phenomenon.

A specific feature that contributes to the potential of prions to act as infectious agents is their high resistance to chemical and biological clearance (Aguzzi and Calella 2009; Prusiner 1998). Prions can resist heating at very high temperatures, inactivation by different classes of enzymes (proteases, nucleases, lipases, etc.), treatments with various kinds of irradiation procedures (UV, ionic, microwaves), traditional sterilization techniques (conventional autoclave, alcohol, disinfectants) and even intestinal digestion. Also important for their pathogenicity is their ability to cross biological membranes, including the intestinal barrier, blood-brain barrier and cellular membranes. This characteristic permits prions to be infectious through a variety of routes, including intra-cerebral, oral, nasal, intra-ocular, intra-peritoneal, etc. (Kimberlin and Walker 1979). Finally, the reported long persistence of prions for many years in the environment (Saunders et al. 2008), added to their well-established ability to attach to surfaces (metal, soil, wood, etc.) and maintaining their infectious processes, contribute to the efficiency of prions acting as infectious agents.

Could AD Be Transmissible?

The common mechanisms responsible for the formation of misfolded aggregates in PMDs, added to the concept that prions are infectious because they act as seeds to propagate the misfolding process, raise the question of whether other PMDs might be transmissible by a prion-like mechanism (Soto et al. 2006). Several attempts have been made to test the potential infectivity of other PMDs. In recent years, and with a better understanding of the molecular basis of prion diseases, the potential transmissibility of several neurodegenerative diseases, such as AD, PD (Desplats et al. 2009; Mougenot et al. 2012), ALS (Grad et al. 2011; Munch et al. 2011) and Huntington's disease (Ren et al. 2009), among others, has been demonstrated. In this section we will focus exclusively on recent studies about the putative transmissibility of Aβ misfolding in AD. However, various excellent reviews have been recently published that discuss the experimental evidence for the transmissibility of other proteins (Brundin et al. 2010; Frost and Diamond 2010; Jucker and Walker 2011; Soto 2012b).

Several studies have been done to understand and evaluate a putative prion-like mechanism involved in the transmission of misfolded Aβ (Table 1). The first approaches to assess AD transmissibility were performed by intra-cerebrally infusing samples from AD patients in several animal models. Among them, non-human

Table 1 Prion transmissibility features tested in vivo for Aβ aggregates

Prion features	Tested for Aβ	Outcome	Reference(s)
In vivo transmission			
	Yes	Positive	Baker et al. (1994), Kane et al. (2000), Meyer-Luehmann et al. (2006), and Morales et al. (2012)
Routes of inoculation			
Intracerebral	Yes	Positive	Baker et al. (1994), Kane et al. (2000), Meyer-Luehmann et al. (2006), and Morales et al. (2012)
Intraperitoneal	Yes	Positive	Eisele et al. (2010)
Intravenous	Yes	Negative	Eisele et al. (2009)
Oral	Yes	Negative	Eisele et al. (2009)
Intranasal	Yes	Negative	Eisele et al. (2009)
Other routes	No	–	
Inter-species transmission			
	No	–	
Existence of strains			
	Yes	Positive[a]	Meyer-Luehmann et al. (2006)
Disease produced by "protein-only" preparations			
	Yes	Positive	Stöhra et al. (2012)
Tested in non-human primates			
	Yes	Negative[b]/Positive	Goudsmit et al. (1980)[b] and Baker et al. (1994)
Reported in humans			
	No	–	

[a]Only one experiment reported (Meyer-Luehmann et al. 2006) and the "strains" were not fully characterized in the manner defined for prion strains
[b]TSEs-like outcome was expected (Goudsmit et al. 1980) and later re-evaluation of these samples using modern techniques led to positive results (Goldgaber, personal communication)

primates and hamsters were injected with human brain extracts (Goudsmit et al. 1980) or blood samples (Manuelidis et al. 1988), respectively. The information from these studies resulted in contradictory results. Experiments from Gadjusek and colleagues showed that intracerebral inoculation of brain homogenates from AD patients in different species of non-human primates failed to transmit the disease (Goudsmit et al. 1980). However, most of these experiments were done before the features of AD aggregates were known, and the animals were expected to show a phenotype similar to that found in prion infection, i.e., spongiform degeneration and obvious clinical signs (motor alterations and ataxia); these abnormalities are not observed in AD. Interestingly, re-evaluation of these tissues with modern antibodies has shown that, indeed, primates injected with AD brain homogenate developed neuropathological features of AD (Goldgaber, personal communication), confirming results obtained by Baker and colleagues in marmoset monkeys.

Intra-cerebral inoculation of AD brain tissue showed the generation of amyloid plaques 6–7 years after inoculation (Baker et al. 1994). Deposits composed of aggregated Aβ were similar to the ones found in the original host, as evaluated by the presence of parenchymal and vascular deposition. Importantly, Aβ deposition was not limited to the injection area, suggesting a diffusion of the newly generated aggregates into other brain structures.

More recent studies by Walker and colleagues took advantage of available transgenic mouse models expressing the human version of APP. These milestone studies demonstrated that β-amyloidogenesis could be induced by unilateral injections of brain homogenates from individuals with AD into tg2576 (APP$_{Swe}$) transgenic mice (Kane et al. 2000; Walker et al. 2002). The mouse model used in this study spontaneously develops human Aβ aggregates due to over-expression and mutation on the human APP sequence. For these experiments, 2-month-old mice received stereotaxic injections into the hippocampus and neocortex and were sacrificed at 5 or 12 months after challenge. Results from these studies showed that brains from AD-injected animals had a characteristic anatomical distribution of Aβ deposits (Kane et al. 2000; Walker et al. 2002). Amyloid plaques appeared scattered preferentially along the hippocampal depression (separation of the dentate gyrus from the subiculum and the CA1) and neocortex (site of injection). Aβ aggregation in the brain of animals after 5 months of infusion was found exclusively in the ipsilateral hemisphere (Kane et al. 2000). Importantly, after 12 months, senile plaques and vascular deposits were located in both hemispheres, but mostly in the half of the brain inoculated with the Aβ-rich tissue (Walker et al. 2002), suggesting a spread of Aβ seeds throughout the brain. The "incubation" time between the inoculation at 2–3 months and sacrifice at 8–9 months of age in tg2576 mouse model has been used in numerous experimental studies. This incubation period allows the evaluation of the Aβ seeding effect before the appearance of endogenously formed plaques.

In further transmission studies using APP23 mice, Mathias Jucker's group analyzed the seeding capabilities of Aβ aggregates in vivo with a similar rationale applied to test infectivity of PrP^{Sc}. These authors showed that the features of induced brain amyloidogenesis depended on both the host and the source of the agent, resembling the strain phenomena in TSEs (Meyer-Luehmann et al. 2006). Using two different transgenic mouse models of AD, APP23 and APP/PS1, they observed dissimilar patterns of Aβ aggregation when brain homogenates from aged mice of one strain were used to seed Aβ deposition on the second one. Interestingly, they also showed that the seeding capability of brain homogenates was reduced or eradicated by different treatments to inactivate aggregated proteins, such as incubation with formic acid, boiling or immunodepletion using anti-Aβ antibodies (Meyer-Luehmann et al. 2006). In addition, this work described the concentration dependence of brain Aβ aggregation when animals were inoculated with 10 and 0.5 % brain homogenates. All these results clearly showed that inoculation of brain homogenates containing preformed Aβ seeds accelerated the deposition of this protein in vivo, using transgenic mice as a model. These results support the hypothesis of a transmissible origin for AD. Additional studies showed that Aβ

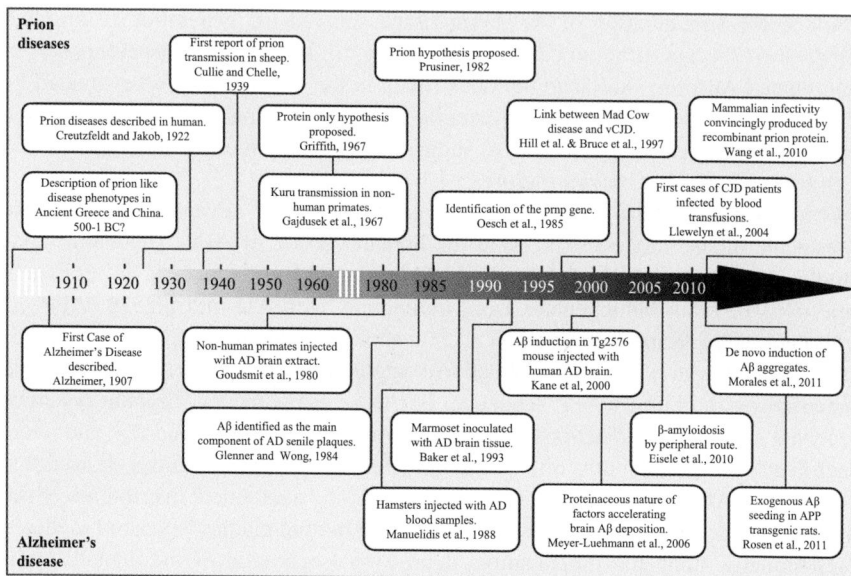

Fig. 2 Timeline for Alzheimer's and prion diseases' transmissibility studies. Although the molecular bases of prion and Aβ aggregation are similar, the possibility that the latter behaves as a transmissible agent has been little explored. Prion disorders have been described for many centuries (McAlister 2005); however, and due to their low incidence in human population, the disease was not described until the early 1920s (Creuztfeldt 1920; Jakob 1921). *CJD* Creutzfeldt-Jakob disease. Transmission studies performed in animals revealed uncommon features for this agent (Gajdusek et al. 1967), leading to the proposal that the infectious agent responsible for these diseases was mainly composed of an abnormal form of a naturally expressed protein (Griffith 1967; Prusiner 1982). In 1984, Aβ was identified as the main component of senile plaques, the hallmark feature of AD. After discouraging experiments performed by the Gajdusek lab, additional Aβ transmissibility studies were performed in non-human primates and transgenic rodents. With the current knowledge collected on prion diseases, understanding the molecular basis of Aβ transmissibility in vivo should become easier

deposition was not induced by some peripheral routes of administration (oral, intravenous, intraocular, intranasal; Eisele et al. 2009). However, steel wires coated with Aβ aggregates were able to seed brain Aβ deposition (Eisele et al. 2009) in a similar way as reported for iatrogenic Creutzfeldt-Jakob disease cases using infected surgical material (Brown et al. 2000) (Fig. 2).

A recent study demonstrated that Aβ deposition could be induced by peripheral administration of seeds (Eisele et al. 2010). Intraperitoneal inoculation of Aβ-rich brain homogenates induced Aβ aggregation as parenchymal and vascular deposits in the brains of APP23 transgenic mice after prolonged incubation times. Although the presence of Aβ aggregates in blood vessels suggested a direct transport of the injected seeds to the brain, further studies are necessary to better understand the cellular and molecular mechanisms that allow transport of Aβ aggregates from the periphery to the brain. Very recent studies have shown that Aβ seeds exist as both protease-resistant and -sensitive species and that soluble Aβ oligomers might be the most potent inducers of amyloidosis (Langer et al. 2011), again similar to what has been shown with prions, where the most infectious prion particles are in the oligomeric range and are protease-sensitive (Silveira et al. 2005).

Despite the importance of these findings, the fact that the experiments were done using transgenic animals that "spontaneously" developed an AD-like pathology at some point in its life raised some questions regarding the similarities with prion diseases. Indeed, it is not possible to distinguish whether seeds from AD brain homogenates produce transmission mimicking an infectious prion agent or only accelerate a process that is invariably set to occur on these genetically programmed animals. Recently, we have shown a de novo induction of Aβ aggregation in vivo (Morales et al. 2012). In this study, we demonstrated that Aβ amyloidogenesis could be induced by injection of AD brain extracts into APP transgenic mice that naturally do not develop Aβ deposition within the course of their life span (>27 months). We found a progressive and time-dependent increase in terms of incidence and Aβ burden after inoculation of AD brain extracts. Animals injected with healthy control brain tissue did not develop Aβ deposits. Aggregates were found not only in the injected area but also far from the injection site. Additionally, we reported a concomitant increase of inflammatory response in the brain of affected animals, suggesting that Aβ deposition induced in these mice was accompanied by other typical AD features (Morales et al. 2012). These results were confirmed by an independent study in which human APP transgenic rats developed Aβ aggregation only after injection of AD brain material (Rosen et al. 2012). In summary, our findings, added to those of Jücker's and Walker's lab, suggest that AD-like pathology (in terms of Aβ aggregation) can be induced by Aβ seeds in a similar way as prions do in TSEs.

Similar data have also been described for the induction of Tau aggregates (de Calignon et al. 2012; Frost et al. 2009; Guo and Lee 2011; Soto 2012a), indicating that not only extracellular aggregates can be seeded in vivo but also those located in the cytoplasm of neurons.

Concluding Remarks

An important unanswered question related to AD involves its etiology, since over 95 % of the cases arise sporadically. Biochemical, genetic, neuropathological and epidemiological data suggest that Aβ misfolding and aggregation are the main events responsible for triggering the pathology. Recent, exciting evidence shows that Aβ aggregation can be induced in vivo following the principles by which prions propagate TSEs. Moreover, similar studies have been done with the majority of neurodegenerative diseases associated with protein misfolding, including PD, ALS and HD (Brundin et al. 2010; Frost and Diamond 2010; Jucker and Walker 2011; Soto 2012b). These findings may open a new opportunity to understand the origin of sporadic AD and may provide new strategies for disease intervention and prevention.

Nevertheless, it remains to be proven whether misfolded aggregates can be transmitted by medically relevant routes of transmission, including blood transfusion, organ transplant, drug extraction from human and animal sources, and oral administration. All those routes have been implicated in the transmission of infectious prions in animals and humans. Even if proven in animal models, further studies are required to explore the possibility of transmission in human beings. Furthermore, it is important to understand the true identity of the agent responsible for seeding. As was seen with prion diseases, the ultimate proof of the protein-only hypothesis (infectious agent generated in vitro by inducing the misfolding of purified proteins) took many years to accomplish (Soto 2011).

Understanding the mechanisms of protein misfolding in vivo is crucial to designing successful diagnostic tests and therapies for the treatment of these devastating diseases. The implications of a transmission process involving AD may have broad impact in public health. However, it is very important to be responsible and cautious when exposing these results, since the extrapolation of experiments in animal models to the disease in humans may not be straightforward.

References

Aguzzi A, Calella AM (2009) Prions: protein aggregation and infectious diseases. Physiol Rev 89:1105–1152

Aguzzi A, Brandner S, Sure U, Ruedi D, Isenmann S (1994) Transgenic and knock-out mice: models of neurological disease. Brain Pathol 4:3–20

Allen B, Ingram E, Takao M, Smith MJ, Jakes R, Virdee K, Yoshida H, Holzer M, Craxton M, Emson PC, Atzori C, Migheli A, Crowther RA, Ghetti B, Spillantini MG, Goedert M (2002) Abundant tau filaments and nonapoptotic neurodegeneration in transgenic mice expressing human P301S tau protein. J Neurosci 22:9340–9351

Alzheimer A (1907) Uber eine eigenartige Erkrankung der Hirnrinde. Allg Zeitschr Psych Gerichtl Med 64:146–148

Arriagada PV, Growdon JH, Hedley-Whyte ET, Hyman BT (1992) Neurofibrillary tangles but not senile plaques parallel duration and severity of Alzheimer's disease. Neurology 42:631–639

Baker HF, Ridley RM, Duchen LW, Crow TJ, Bruton CJ (1994) Induction of beta (A4)-amyloid in primates by injection of Alzheimer's disease brain homogenate. Comparison with transmission of spongiform encephalopathy. Mol Neurobiol 8:25–39

Ballard C, Gauthier S, Corbett A, Brayne C, Aarsland D, Jones E (2011) Alzheimer's disease. Lancet 377:1019–1031

Blake CC, Serpell LC, Sunde M, Sandgren O, Lundgren E (1996) A molecular model of the amyloid fibril. Ciba Found Symp 199:6–15

Borchelt DR, Ratovitski T, van Lare J, Lee MK, Gonzales V, Jenkins NA, Copeland NG, Price DL, Sisodia SS (1997) Accelerated amyloid deposition in the brains of transgenic mice coexpressing mutant presenilin 1 and amyloid precursor proteins. Neuron 19:939–945

Braak H, Braak E (1991) Neuropathological staging of Alzheimer-related changes. Acta Neuropathol (Berl) 82:239–259

Brandt R, Hundelt M, Shahani N (2005) Tau alteration and neuronal degeneration in tauopathies: mechanisms and models. Biochim Biophys Acta 1739:331–354

Brayne C (2007) The elephant in the room – healthy brains in later life, epidemiology and public health. Nat Rev Neurosci 8:233–239

Brendza RP, Bales KR, Paul SM, Holtzman DM (2002) Role of apoE/Abeta interactions in Alzheimer's disease: insights from transgenic mouse models. Mol Psychiatry 7:132–135

Brookmeyer R, Johnson E, Ziegler-Graham K, Arrighi HM (2007) Forecasting the global burden of Alzheimer's disease. Alzheimers Dement 3:186–191

Brown P, Preece M, Brandel JP, Sato T, McShane L, Zerr I, Fletcher A, Will RG, Pocchiari M, Cashman NR, Preece M, Brandel JP, Sato T, McShane L, Zerr I, Fletcher A, Will RG, Pocchiari M, Cashman NR, d'Aignaux JH, Cervenáková L, Fradkin J, Schonberger LB, Collins SJ (2000) Iatrogenic Creutzfeldt-Jakob disease at the millennium. Neurology 55:1075–1081

Brundin P, Melki R, Kopito R (2010) Prion-like transmission of protein aggregates in neurodegenerative diseases. Nat Rev Mol Cell Biol 11:301–307

Calhoun ME, Wiederhold KH, Abramowski D, Phinney AL, Probst A, Sturchler-Pierrat C, Staufenbiel M, Sommer B, Jucker M (1998) Neuron loss in APP transgenic mice. Nature 395:755–756

Calhoun ME, Burgermeister P, Phinney AL, Stalder M, Tolnay M, Wiederhold KH, Abramowski D, Sturchler-Pierrat C, Sommer B, Staufenbiel M, Jucker M (1999) Neuronal overexpression of mutant amyloid precursor protein results in prominent deposition of cerebrovascular amyloid. Proc Natl Acad Sci USA 96:14088–14093

Chapman PF, Falinska AM, Knevett SG, Ramsay MF (2001) Genes, models and Alzheimer's disease. Trends Genet 17:254–261

Cheon M, Favrin G, Chang I, Dobson CM, Vendruscolo M (2008) Calculation of the free energy barriers in the oligomerisation of Abeta peptide fragments. Front Biosci 13:5614–5622

Cleary JP, Walsh DM, Hofmeister JJ, Shankar GM, Kuskowski MA, Selkoe DJ, Ashe KH (2004) Natural oligomers of the amyloid-beta protein specifically disrupt cognitive function. Nat Neurosci 8:79–84

Clinton LK, Blurton-Jones M, Myczek K, Trojanowski JQ, LaFerla FM (2010) Synergistic interactions between Abeta, tau, and alpha-synuclein: acceleration of neuropathology and cognitive decline. J Neurosci 30:7281–7289

Collinge J (2001) Prion diseases of humans and animals: their causes and molecular basis. Annu Rev Neurosci 24:519–550

Creuztfeldt H (1920) Über eine eigenartige herdförmige Erkrankung des Zentralnervensystems. Z Ges Neurol Psychiat 57:1–18

de Calignon A, Polydoro M, Suarez-Calvet M, William C, Adamowicz DH, Kopeikina KJ, Pitstick R, Sahara N, Ashe KH, Carlson GA, Spires-Jones TL, Hyman BT (2012) Propagation of tau pathology in a model of early Alzheimer's disease. Neuron 73:685–697

Desplats P, Lee HJ, Bae EJ, Patrick C, Rockenstein E, Crews L, Spencer B, Masliah E, Lee SJ (2009) Inclusion formation and neuronal cell death through neuron-to-neuron transmission of alpha-synuclein. Proc Natl Acad Sci USA 106:13010–13015

Duff K, Eckman C, Zehr C, Yu X, Prada CM, Perez-tur J, Hutton M, Buee L, Harigaya Y, Yager D, Morgan D, Gordon MN, Holcomb L, Refolo L, Zenk B, Hardy J, Younkin S (1996) Increased amyloid-beta42(43) in brains of mice expressing mutant presenilin 1. Nature 383:710–713

Eisele YS, Bolmont T, Heikenwalder M, Langer F, Jacobson LH, Yan ZX, Roth K, Aguzzi A, Staufenbiel M, Walker LC, Jucker M (2009) Induction of cerebral beta-amyloidosis: intracerebral versus systemic Abeta inoculation. Proc Natl Acad Sci USA 106:12926–12931

Eisele YS, Obermuller U, Heilbronner G, Baumann F, Kaeser SA, Wolburg H, Walker LC, Staufenbiel M, Heikenwalder M, Jucker M (2010) Peripherally applied A{beta}-containing inoculates induce cerebral beta-amyloidosis. Science 330:980–982

Frost B, Diamond MI (2010) Prion-like mechanisms in neurodegenerative diseases. Nat Rev Neurosci 11:155–159

Frost B, Jacks RL, Diamond MI (2009) Propagation of tau misfolding from the outside to the inside of a cell. J Biol Chem 284:12845–12852

Gajdusek DC, Gibbs CJ Jr, Alpers M (1967) Transmission and passage of experimental "kuru" to chimpanzees. Science 155:212–214

Games D, Adams D, Alessandrini R, Barbour R, Berthelette P, Blackwell C, Carr T, Clemens J, Donaldson T, Gillespie Guido T, Hagopian S, Johnson-Wood K, Khan K, Lee M, Leibowitz P, Lieberburg I, Little S, Masliah E, McConlogue L, Montoya-Zavala M, Mucke L, Paganini L, Penniman E, Power M, Schenk D, Seubert P, Snyder B, Soriano F, Tan H, Vitale J, Wadsworth S, Wolozin B, Zhao J (1995) Alzheimer-type neuropathology in transgenic mice overexpressing V717F beta-amyloid precursor protein. Nature 373:523–527

Glabe CG (2006) Common mechanisms of amyloid oligomer pathogenesis in degenerative disease. Neurobiol Aging 27:570–575

Glenner GG, Wong CW (1984) Alzheimer's disease: initial report of the purification and characterization of a novel cerebrovascular amyloid protein. Biochem Biophys Res Commun 120:885–890

Goate A, Chartier-Harlin MC, Mullan M, Brown J, Crawford F, Fidani L, Giuffra L, Haynes A, Irving N, James L, Mant R, Newton P, Rooke K, Roques P, Talbot C, Pericak-Vance M, Roses A, Williamson R, Rossor M, Owen M, Hardy J (1991) Segregation of a missense mutation in the amyloid precursor protein gene with familial Alzheimer's disease. Nature 349:704–706

Goudsmit J, Morrow CH, Asher DM, Yanagihara RT, Masters CL, Gibbs CJ Jr, Gajdusek DC (1980) Evidence for and against the transmissibility of Alzheimer disease. Neurology 30:945–950

Grad LI, Guest WC, Yanai A, Pokrishevsky E, O'Neill MA, Gibbs E, Semenchenko V, Yousefi M, Wishart DS, Plotkin SS, Cashman NR (2011) Intermolecular transmission of superoxide dismutase 1 misfolding in living cells. Proc Natl Acad Sci USA 108:16398–16403

Griffith JS (1967) Self-replication and scrapie. Nature 215:1043–1044

Grundke-Iqbal I, Iqbal K, Quinlan M, Tung YC, Zaidi MS, Wisniewski HM (1986) Microtubule-associated protein tau A component of Alzheimer paired helical filaments. J Biol Chem 261:6084–6089

Guo JL, Lee VM (2011) Seeding of normal Tau by pathological Tau conformers drives pathogenesis of Alzheimer-like tangles. J Biol Chem 286:15317–15331

Haass C, Selkoe DJ (2007) Soluble protein oligomers in neurodegeneration: lessons from the Alzheimer's amyloid beta-peptide. Nat Rev Mol Cell Biol 8:101–112

Hardy J (1992) An 'anatomical cascade hypothesis' for Alzheimer's disease. Trends Neurosci 15:200–201

Hardy J, Gwinn-Hardy K (1998) Genetic classification of primary neurodegenerative disease. Science 282:1075–1079

Hsiao K, Chapman P, Nilsen S, Eckman C, Harigaya Y, Younkin S, Yang F, Cole G (1996) Correlative memory deficits, Abeta elevation, and amyloid plaques in transgenic mice. Science 274:99–102

Jakob A (1921) Über eigenartige Erkrankungen des Zentralnervensystems mit bemerkenswertenanatomischen Befunde (Spastische Pseudosklerose-Encephalomyelopathie mit disseminiertenDegenerationsherden). Z Ges Neurol Psychiat 64:147–228

Jarrett JT, Lansbury PT Jr (1993) Seeding "one-dimensional crystallization" of amyloid: a pathogenic mechanism in Alzheimer's disease and scrapie? Cell 73:1055–1058

Jucker M, Walker LC (2011) Pathogenic protein seeding in Alzheimer disease and other neurodegenerative disorders. Ann Neurol 70:532–540

Kane MD, Lipinski WJ, Callahan MJ, Bian F, Durham RA, Schwarz RD, Roher AE, Walker LC (2000) Evidence for seeding of beta -amyloid by intracerebral infusion of Alzheimer brain extracts in beta-amyloid precursor protein-transgenic mice. J Neurosci 20:3606–3611

Kawarabayashi T, Younkin LH, Saido TC, Shoji M, Ashe KH, Younkin SG (2001) Age-dependent changes in brain, CSF, and plasma amyloid (beta) protein in the Tg2576 transgenic mouse model of Alzheimer's disease. J Neurosci 21:372–381

Kimberlin RH, Walker CA (1979) Pathogenesis of mouse scrapie: dynamics of agent replication in spleen, spinal cord and brain after infection by different routes. J Comp Pathol 89:551–562

Klein WL, Krafft GA, Finch CE (2001) Targeting small Abeta oligomers: the solution to an Alzheimer's disease conundrum? Trends Neurosci 24:219–224

Koo EH, Lansbury PT Jr, Kelly JW (1999) Amyloid diseases: abnormal protein aggregation in neurodegeneration. Proc Natl Acad Sci USA 96:9989–9990

Langer F, Eisele YS, Fritschi SK, Staufenbiel M, Walker LC, Jucker M (2011) Soluble Abeta seeds are potent inducers of cerebral beta-amyloid deposition. J Neurosci 31:14488–14495

Lesne S, Koh MT, Kotilinek L, Kayed R, Glabe CG, Yang A, Gallagher M, Ashe KH (2006) A specific amyloid-beta protein assembly in the brain impairs memory. Nature 440:352–357

Levy-Lahad E, Wasco W, Poorkaj P, Romano DM, Oshima J, Pettingell WH, Yu CE, Jondro PD, Schmidt SD, Wang K, Crowley AC, Fu YH, Guenette SY, Galas D, Nemens E, Wijsman EM, Bird TD, Schellenberg GD, Tanzi RE (1995) Candidate gene for the chromosome 1 familial Alzheimer's disease locus. Science 269:973–977

Luheshi LM, Crowther DC, Dobson CM (2008) Protein misfolding and disease: from the test tube to the organism. Curr Opin Chem Biol 12:25–31

Manuelidis EE, de Figueiredo JM, Kim JH, Fritch WW, Manuelidis L (1988) Transmission studies from blood of Alzheimer disease patients and healthy relatives. Proc Natl Acad Sci USA 85:4898–4901

McAlister V (2005) Sacred disease of our times: failure of the infectious disease model of spongiform encephalopathy. Clin Invest Med 28:101–104

McGeer PL, McGeer EG (2007) NSAIDs and Alzheimer disease: epidemiological, animal model and clinical studies. Neurobiol Aging 28:639–647

McGowan E, Eriksen J, Hutton M (2006) A decade of modeling Alzheimer's disease in transgenic mice. Trends Genet 22:281–289

Meyer-Luehmann M, Coomaraswamy J, Bolmont T, Kaeser S, Schaefer C, Kilger E, Neuenschwander A, Abramowski D, Frey P, Jato AL, Vigouret JM, Paganetti P, Walsh DM, Mathews PM, Ghiso J, Staufenbiel M, Walker LC, Jucker M (2006) Exogenous induction of cerebral beta-amyloidogenesis is governed by agent and host. Science 313:1781–1784

Morales R, Green KM, Soto C (2009) Cross currents in protein misfolding disorders: interactions and therapy. CNS Neurol Disord Drug Targets 8:363–371

Morales R, Duran-Aniotz C, Castilla J, Estrada LD, Soto C (2012) De novo induction of amyloid-beta deposition in vivo. Mol Psychiatry 17(12):1347–1353. doi:10.1038/mp. 2011.120

Mougenot AL, Nicot S, Bencsik A, Morignat E, Verchere J, Lakhdar L, Legastelois S, Baron T (2012) Prion-like acceleration of a synucleinopathy in a transgenic mouse model. Neurobiol Aging 33:2225–2228

Munch C, O'Brien J, Bertolotti A (2011) Prion-like propagation of mutant superoxide dismutase-1 misfolding in neuronal cells. Proc Natl Acad Sci USA 108:3548–3553

Oakley H, Cole SL, Logan S, Maus E, Shao P, Craft J, Guillozet-Bongaarts A, Ohno M, Disterhoft J, Van EL, Berry R, Vassar R (2006) Intraneuronal beta-amyloid aggregates, neurodegeneration, and neuron loss in transgenic mice with five familial Alzheimer's disease mutations: potential factors in amyloid plaque formation. J Neurosci 26:10129–10140

Oddo S, Caccamo A, Shepherd JD, Murphy MP, Golde TE, Kayed R, Metherate R, Mattson MP, Akbari Y, LaFerla FM (2003) Triple-transgenic model of Alzheimer's disease with plaques and tangles: intracellular Abeta and synaptic dysfunction. Neuron 39:409–421

Price JL, McKeel DW Jr, Buckles VD, Roe CM, Xiong C, Grundman M, Hansen LA, Petersen RC, Parisi JE, Dickson DW, Smith CD, Davis DG, Schmitt FA, Markesbery WR, Kaye J, Kurlan R, Hulette C, Kurland BF, Higdon R, Kukull W, Morris JC (2009) Neuropathology of nondemented aging: presumptive evidence for preclinical Alzheimer disease. Neurobiol Aging 30:1026–1036

Prusiner SB (1982) Novel proteinaceous infectious particles cause scrapie. Science 216:136–144

Prusiner SB (1998) Prions. Proc Natl Acad Sci USA 95:13363–13383

Reed MN, Hofmeister JJ, Jungbauer L, Welzel AT, Yu C, Sherman MA, Lesne S, LaDu MJ, Walsh DM, Ashe KH, Cleary JP (2011) Cognitive effects of cell-derived and synthetically derived Abeta oligomers. Neurobiol Aging 32:1784–1794

Reitz C, Brayne C, Mayeux R (2011) Epidemiology of Alzheimer disease. Nat Rev Neurol 7:137–152

Ren PH, Lauckner JE, Kachirskaia I, Heuser JE, Melki R, Kopito RR (2009) Cytoplasmic penetration and persistent infection of mammalian cells by polyglutamine aggregates. Nat Cell Biol 11:219–225

Rockenstein E, Crews L, Masliah E (2007) Transgenic animal models of neurodegenerative diseases and their application to treatment development. Adv Drug Deliv Rev 59:1093–1102

Rosen RF, Fritz JJ, Dooyema J, Cintron AF, Hamaguchi T, Lah JJ, LeVine H III, Jucker M, Walker LC (2012) Exogenous seeding of cerebral beta-amyloid deposition in betaAPP-transgenic rats. J Neurochem 120:660–666

Roses AD, Saunders AM (1994) APOE is a major susceptibility gene for Alzheimer's disease. Curr Opin Biotechnol 5:663–667

Saunders SE, Bartelt-Hunt SL, Bartz JC (2008) Prions in the environment: occurrence, fate and mitigation. Prion 2:162–169

Schellenberg GD, Bird TD, Wijsman EM, Orr HT, Anderson L, Nemens E, White JA, Bonnycastle L, Weber JL, Alonso ME, Potter H, Heston LL, Martin GM (1992) Genetic linkage evidence for a familial Alzheimer's disease locus on chromosome 14. Science 258:668–671

Serrano-Pozo A, Frosch MP, Masliah E, Hyman BT (2011) Neuropathological alterations in Alzheimer disease. Cold Spring Harb Perspect Med 1:a006189

Shankar GM, Bloodgood BL, Townsend M, Walsh DM, Selkoe DJ, Sabatini BL (2007) Natural oligomers of the Alzheimer amyloid-beta protein induce reversible synapse loss by modulating an NMDA-type glutamate receptor-dependent signaling pathway. J Neurosci 27:2866–2875

Sigurdson CJ, Aguzzi A (2006) Chronic wasting disease. Biochim Biophys Acta 1772:610–618

Silveira JR, Raymond GJ, Hughson AG, Race RE, Sim VL, Hayes SF, Caughey B (2005) The most infectious prion protein particles. Nature 437:257–261

Soto C (2003) Unfolding the role of protein misfolding in neurodegenerative diseases. Nat Rev Neurosci 4:49–60

Soto C (2004) Diagnosing prion diseases: needs, challenges and hopes. Nat Rev Microbiol 2:809–819

Soto C (2011) Prion hypothesis: the end of the controversy? Trends Biochem Sci 36:151–158

Soto C (2012a) In vivo spreading of tau pathology. Neuron 73:621–623

Soto C (2012b) Transmissible proteins: expanding the prion heresy. Cell 149:968–977

Soto C, Estrada L, Castilla J (2006) Amyloids, prions and the inherent infectious nature of misfolded protein aggregates. Trends Biochem Sci 31:150–155

Stöhr J, Watts JC, Mensinger ZL, Oehler A, Grillo SK, DeArmond SJ, Prusiner SB, Giles K (2012) Purified and synthetic Alzheimer's amyloid beta (Aβ) prions. Proc Natl Acad Sci USA 109(27):11025–11030

Sturchler-Pierrat C, Abramowski D, Duke M, Wiederhold KH, Mistl C, Rothacher S, Ledermann B, Burki K, Frey P, Paganetti PA, Waridel C, Calhoun ME, Jucker M, Probst A, Staufenbiel M, Sommer B (1997) Two amyloid precursor protein transgenic mouse models with Alzheimer disease-like pathology. Proc Natl Acad Sci USA 94:13287–13292

Thomas PJ, Qu B-H, Pedersen PL (1995) Defective protein folding as a basis of human disease. Trends Biochem Sci 20:456–459

Townsend M, Shankar GM, Mehta T, Walsh DM, Selkoe DJ (2006) Effects of secreted oligomers of amyloid beta-protein on hippocampal synaptic plasticity: a potent role for trimers. J Physiol 572:477–492

Vossel KA, Zhang K, Brodbeck J, Daub AC, Sharma P, Finkbeiner S, Cui B, Mucke L (2010) Tau reduction prevents Abeta-induced defects in axonal transport. Science 330:198

Walker LC, Callahan MJ, Bian F, Durham RA, Roher AE, Lipinski WJ (2002) Exogenous induction of cerebral beta-amyloidosis in betaAPP-transgenic mice. Peptides 23:1241–1247

Walsh DM, Selkoe DJ (2007) A beta oligomers – a decade of discovery. J Neurochem 101:1172–1184

Walsh DM, Klyubin I, Fadeeva JV, Cullen WK, Anwyl R, Wolfe MS, Rowan MJ, Selkoe DJ (2002) Naturally secreted oligomers of amyloid beta protein potently inhibit hippocampal long-term potentiation in vivo. Nature 416:535–539

Prion-Like Properties of Assembled Tau Protein

Florence Clavaguera, Markus Tolnay, and Michel Goedert

Abstract The soluble microtubule-associated protein tau becomes hyperphosphorylated, insoluble and filamentous in a number of neurodegenerative diseases collectively referred to as tauopathies. In Alzheimer's disease (AD), tau pathology develops in a stereotypical manner, with the first lesions appearing in the locus coeruleus and the transentorhinal cortex, from where they appear to spread to the entorhinal cortex, the hippocampus and the neocortex. The staging of tau pathology has also been described in argyrophilic grain disease (AGD), where tau lesions spread stereotypically throughout the limbic system. The isoform composition and morphology of tau filaments differ between diseases, suggesting the possible existence of different tau strains, reminiscent of prion strains. Prion diseases result from the misfolding of the cellular prion protein that can occur sporadically, as the result of dominantly inherited mutations or following infection. Recent experimental work has shown that prion-like mechanisms are also at work in the tauopathies.

Prions and Tauopathies

Prions give rise to severe neurodegenerative diseases, which are mostly sporadic, but which can also be inherited or acquired through infection (Prusiner 1982; Colby and Prusiner 2011). A conformational change in the prion protein underlies these diseases and turns the normal cellular PrP^C into the pathogenic PrP^{Sc}, which is rich in β-sheets (Caughey and Raymond 1991). PrP^{Sc} recruits PrP^C and converts it into

F. Clavaguera • M. Tolnay
Institute of Pathology, University of Basel, Schönbeinstrasse 40, 4031 Basel, Switzerland

M. Goedert (✉)
MRC Laboratory of Molecular Biology, Hills Road, Cambridge CB2 0QH, UK
e-mail: mg@mrc-lmb.cam.ac.uk

M. Jucker and Y. Christen (eds.), *Proteopathic Seeds and Neurodegenerative Diseases*,
Research and Perspectives in Alzheimer's Disease,
DOI 10.1007/978-3-642-35491-5_7, © Springer-Verlag Berlin Heidelberg 2013

Table 1 Diseases with tau inclusions

Alzheimer's disease
Amyotrophic lateral sclerosis/parkinsonism-dementia complex
Argyrophilic grain disease
Chronic traumatic encephalopathy
Corticobasal degeneration
Diffuse neurofibrillary tangles with calcification
Down's syndrome
Familial British dementia
Familial Danish dementia
Frontal variant of Alzheimer's disease
Frontotemporal dementia and parkinsonism linked to chromosome 17 caused by *MAPT* mutations
Gerstmann-Sträussler-Scheinker disease
Guadeloupean parkinsonism
Myotonic dystrophy
Neurodegeneration with brain iron accumulation
Niemann-Pick disease, type C
Non-Guamanian motor neuron disease with neurofibrillary tangles
Pick's disease
Postencephalitic parkinsonism
Prion protein cerebral amyloid angiopathy
Progressive subcortical gliosis
Progressive supranuclear palsy
SLC9A6-related mental retardation
Subacute sclerosing panencephalitis
Tangle-only dementia
White matter tauopathy with globular glial inclusions

the pathogenic form through a process of nucleated polymerization. PrP^{Sc} can adopt several different conformations that encode distinct prion strains that, in turn, give rise to diseases with unique incubation periods, pathologies and clinical manifestations (Dickinson and Meikle 1969; Tanaka et al. 2006; Colby et al. 2009).

Alzheimer's disease (AD) is the most common form of dementia and affects millions of people worldwide. All six brain tau isoforms in a hyperphosphorylated state make up the intracellular filamentous deposits that take the form of neurofibrillary tangles (NFTs), neuropil threads (NTs) and dystrophic neurites. The latter are associated with extracellular Aβ plaques, which constitute the other defining neuropathological characteristic of AD (Goedert and Spillantini 2006). The six tau isoforms, which are produced from a single gene (*MAPT*) through alternative mRNA splicing, can be divided into two groups of three isoforms each: those with three tandem repeats (3R) and those with four tandem repeats (4R: Goedert et al. 1989). Abundant pathological tau inclusions in the absence of Aβ deposits characterize a large number of additional neurodegenerative diseases (Table 1). Although they can have some lesions in common, including brain atrophy, neuronal loss, gliosis, spongiosis, ballooned neurons and abnormal glial cells, tauopathies also show disease-specific changes (Lee et al. 2001).

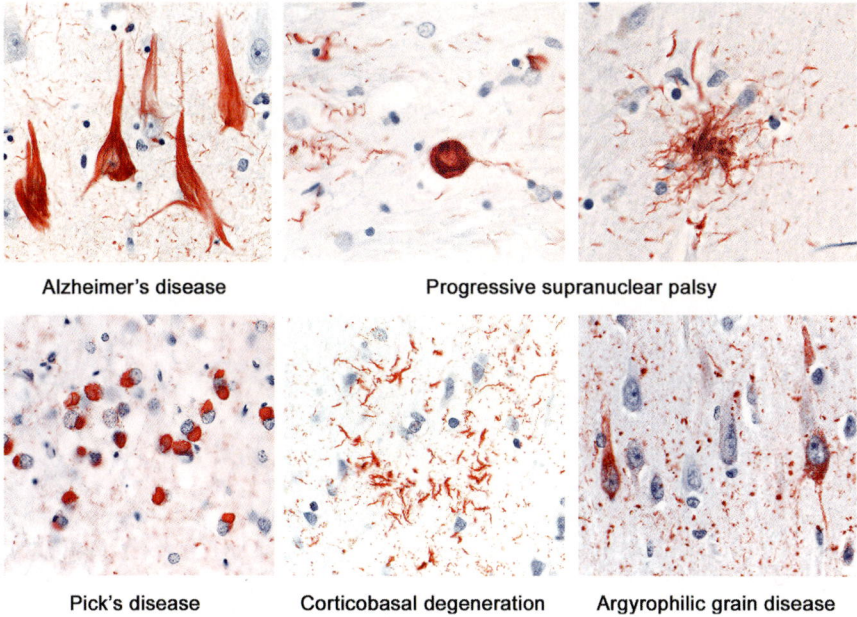

Fig. 1 Different types of tau pathology in human tauopathies: immunohistochemistry with phosphorylation-dependent anti-tau antibody AT8. From *left* to *right*, *upper panel*: typical flame-shaped neurofibrillary tangles in Alzheimer's disease (AD); globose-type neurofibrillary tangles and tufted astrocytes in progressive supranuclear palsy (PSP). *Lower panel*: Pick bodies in Pick's disease; astrocytic plaques in corticobasal degeneration (CBD); argyrophilic grains and pre-tangle neurons in argyrophilic grain disease (AGD). In AD, tau filaments are made of all six human brain tau isoforms (3R + 4R). In Pick's disease, tau filaments consist of three tau isoforms with 3R each. In PSP, CBD and AGD, tau filaments are made of three tau isoforms with 4R each

In 1998, *MAPT* mutations were shown to cause cases of frontotemporal dementia and parkinsonism linked to chromosome 17 (FTDP-17T), a condition associated with abundant tau-positive inclusions, demonstrating that dysfunction of tau protein is sufficient to cause neurodegeneration and dementia (Hutton et al. 1998; Poorkaj et al. 1998; Spillantini et al. 1998). In FTDP-17T, as in sporadic tauopathies (Fig. 1), tau aggregates are composed of 3R tau, 4R tau or a combination of 3R and 4R tau isoforms (Crowther and Goedert 2000; Goedert and Spillantini 2006).

In the process of AD, filamentous inclusions made of all six human brain tau isoforms develop in a stereotypical manner. This pattern forms the basis of the so-called 'Braak stages' (Braak and Braak 1991; Braak and Del Tredici 2011). During stages I and II, which are not associated with dementia, tau lesions develop in the locus coeruleus and the transentorhinal cortex. The more pronounced involvement of the transentorhinal cortex and the formation of NFTs in the entorhinal cortex and the hippocampus are characteristic of stages III–IV. The degree of neuronal damage in stages III–IV may lead to the appearance of the first clinical symptoms. In stages V and VI, abundant NFTs are present in isocortical association areas.

Patients with Braak stages V and VI are severely demented and meet the clinical criteria of AD. Stereotypical spatiotemporal spreading of tau inclusions has also been described in AGD, a 4R tauopathy (Saito et al. 2004). The earliest changes are restricted to the ambient gyrus (stage I), from where the pathological process extends to the anterior and posterior medial temporal lobe (stage II), followed by the septum, insular cortex and anterior cingulate gyrus (stage III). Stage III is characteristic of patients with a clinical diagnosis of dementia.

Experimental Transmission of Tauopathy

We have studied the experimental induction and propagation of tau pathology using transgenic mouse lines ALZ17 and P301S tau (Clavaguera et al. 2009). ALZ17 mice express the longest human brain tau isoform (441 amino acids) but do not develop filamentous aggregates (Probst et al. 2000). By contrast, P301S tau mice, which express the 383 amino acid human tau isoform with the P301S mutation that causes FTDP-17T (Bugiani et al. 1999), develop abundant filamentous tau inclusions (Allen et al. 2002). To determine whether the aggregation of tau protein can be induced *in vivo*, we injected brainstem homogenates from 5-month-old P301S tau mice into the hippocampus and the overlying cerebral cortex of 3-month-old ALZ17 mice (Fig. 2). This procedure induced the assembly of wild-type human tau into filaments, not only in neurons but also in oligodendrocytes. Strikingly, the induction of filamentous tau was not restricted to the injection sites but progressed over time to both neighboring and more remote, but anatomically connected, brain regions. In addition, filamentous tau was induced in different cell types, since we observed NFTs, NTs and oligodendroglial coiled bodies similar to those found in the human tauopathies (Fig. 3). Injection of brain extracts from P301S tau mice that were devoid of tau, or contained only soluble or insoluble tau, showed that it was insoluble tau that induced aggregation (Clavaguera et al. 2009). Recent studies have confirmed and extended these findings (Liu et al. 2012; De Calignon et al. 2012).

To investigate whether tau pathology can propagate along the entorhinal cortex/hippocampal pathway, expression of human mutant P301L tau was induced in a restricted fashion in the entorhinal cortex. The brain distribution of tau pathology was then analyzed, with a special emphasis on areas that are connected to the entorhinal cortex. The mice developed an age-dependent accumulation of tau pathology in neurons of the entorhinal cortex that expressed the transgene. Neurons in the hippocampal formation developed tau inclusions several months after the first inclusions had formed in the entorhinal cortex. Both studies appear rule out the possibility that the tau pathology observed outside the entorhinal cortex resulted from leaky expression of the transgene. They demonstrated that human tau was present in neurons that did not express detectable levels of the transgene, favoring a mechanism based on the neuron-to-neuron propagation of assembled tau in

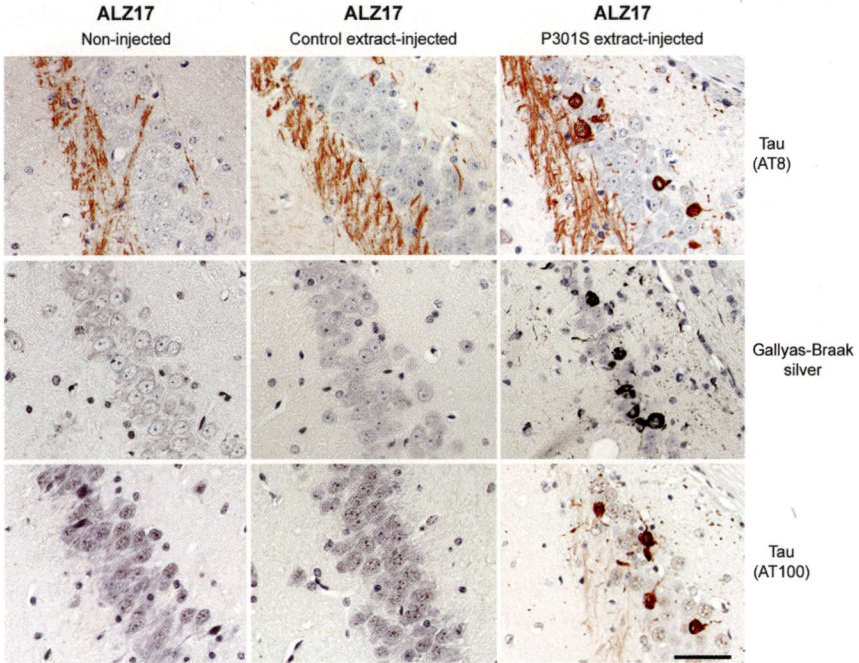

Fig. 2 Induction of filamentous tau pathology in ALZ17 brains injected with brain extract from mice transgenic for human P301S tau: staining of the hippocampal CA3 region from 18-month-old ALZ17 mice with anti-tau antibody AT8, Gallyas-Braak silver and phosphorylation-dependent anti-tau antibody AT100. Non-injected (*left*), 15 months after injection with brain extract from non-transgenic control mice (*middle*) and 15 months after injection with brain extract from 6-month-old mice transgenic for human P301S tau (*right*). The sections were counterstained with haematoxylin. *Scale bar*, 50 μm (same magnification in all panels)

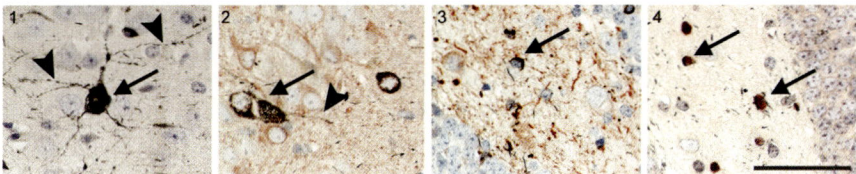

Fig. 3 Different types of filamentous tau pathology in ALZ17 brains injected with brain extract from mice transgenic for human P301S tau: (*1*) Neurofibrillary tangle (*arrow*) and neuropil threads (*arrowheads*) visualized by silver staining; (*2, 3*) double staining with silver and AT8 (*red*) shows neurofibrillary tangles (*2, arrow*), neuropil threads (*2, arrowhead*) and coiled bodies (*3, arrow*). (*4*) Double staining of coiled bodies (*arrows*) with silver and antibody Olig2 (*red*). The sections were counterstained with haematoxylin. *Scale bar*, 50 μm (same magnification in all panels)

the absence of tau overexpression outside the entorhinal cortex. Mouse tau co-aggregated with human tau, indicative of seeding across species.

In vitro studies have also shown the induction and propagation of tau misfolding. When added to non-neuronal cells expressing soluble tau, tau aggregates induced misfolding of intracellular tau (Frost et al. 2009a). The aggregates co-localized with dextran, a marker of fluid phase endocytosis, but not with cholera toxin B, a marker of lipid rafts, demonstrating the relevance of endocytic pathways. Aggregated intracellular tau seeded further aggregation and could transfer between cells. The seed-dependent polymerization of tau was also demonstrated when lipofection was used to introduce amyloid seeds into cultured cells (Nonaka et al. 2010). In this model, 3R or 4R tau was transiently expressed in neuroblastoma cells. Filaments made of 3R tau seeded aggregation in 3R tau but not in 4R tau-expressing cells. Conversely, 4R tau filaments induced aggregation of 4R tau but not 3R tau, suggesting that the specific assembly into amyloid filaments in the presence of seeds derived from tau protein with the same number of repeats. Separate in vitro work has demonstrated the templated transmission of the conformational properties of assembled recombinant tau. Filaments made of either wild-type or mutant tau adopted distinct conformations that were maintained via a templated conformational change (Frost et al. 2009b).

It is well established that tau filaments form through a nucleation-dependent polymerization mechanism in vitro (Friedhoff et al. 1998). In cells, the induction of insoluble tau aggregates also requires seeding (Guo and Lee 2011). Filaments were generated from either myc-tagged full-length human tau or tau consisting of only 4R. When these filaments were transduced into cells expressing the longest human brain tau isoform, intracellular tau aggregates of various morphologies formed. The seeding of endogenous wild-type tau was more efficient when filaments were prepared from human mutant tau than when they were made from human wild-type tau. Furthermore, cells expressing soluble human mutant tau and transduced with filaments made of the same human mutant tau showed the largest recruitment of soluble tau. The aggregates were thioflavin S-positive, indicating the presence of β-pleated sheets. To investigate spontaneous uptake, cells were incubated with filaments in the absence of a protein delivery reagent. Tau aggregates formed in about 10 % of cells. Blocking endocytosis reduced the percentage of aggregate-containing cells, whereas favoring adsorptive endocytosis increased the number of cells with tau aggregates. These findings indicate that the spontaneous entry of tau filaments occurred through endocytosis.

Conclusion

The mechanisms underlying the stereotypical spreading of tau pathology in AD and AGD remain to be discovered. For many years, it was believed that tau pathology developed cell-autonomously. In the rare cases with dominantly inherited pathogenic *MAPT* mutations, this explanation appears plausible, because the widely expressed human mutant tau tends to have an increased propensity to aggregate. However, for the much more common cases of sporadic disease, it is difficult to envisage how the same specific misfolding of wild-type tau could occur independently tens of thousands of times. The work described here now shows that tau aggregates can transfer between cells and function as seeds that nucleate more aggregation. It is therefore conceivable that one or a small number of tau misfolding events are capable of initiating what may result in clinical disease many years later. In our study, ALZ17 mice injected with brainstem extract from mice transgenic for human mutant P301S tau developed a robust filamentous tau pathology that increased over time and appeared in regions with neuronal connections to the injection sites, reminiscent of the Braak stages (Clavaguera et al. 2009). However, no obvious signs of neurodegeneration were detected in ALZ17 mice injected with P301S tau brain extract, in contrast to the P301S tau parent line, suggesting that the molecular tau species responsible for transmission and toxicity are different. Parallel studies have demonstrated templated transmission of the conformational properties of assembled recombinant tau in vitro, suggesting the existence of tau strains and further emphasizing similarities between prion diseases and tauopathies (Frost et al. 2009a).

Tauopathies are heterogeneous diseases that present with clinical and phenotypic differences in which tau aggregates show polymorphic conformations. Distinct tau strains may therefore explain the pathogenic and phenotypic variations between tauopathies. In prion diseases, a strong link exists between protein conformation and clinical phenotype (Collinge and Clarke 2007). In vitro studies have shown that tau aggregates can be released and taken up by neighboring cells through endocytosis. Similar mechanisms may operate in AD (many tau aggregates accumulate in the extracellular space as ghost tangles following the death of tangle-bearing neurons), emphasizing the potential of immunotherapeutic approaches for human tauopathies (Asuni et al. 2007). The prion-like induction and propagation of protein aggregation are general principles that underlie the neurodegenerative proteinopathies. They are bound to guide the development of mechanism-based therapies (Goedert et al. 2010).

References

Allen B, Ingram E, Takao M, Smith MJ, Jakes R, Virdee K, Yoshida H, Holzer M, Craxton M, Emson PC, Atzori C, Migheli A, Crowther RA, Ghetti B, Spillantini MG, Goedert M (2002) Abundant tau filaments and nonapoptotic neurodegeneration in transgenic mice expressing human P301S tau protein. J Neurosci 22:9340–9351

Asuni AA, Boutajangout A, Quartermain D, Sigurdsson EM (2007) Immunotherapy targeting pathological tau conformers in a tangle mouse model reduces brain pathology with associated functional improvements. J Neurosci 27:9115–9129

Braak H, Braak E (1991) Neuropathological stageing of Alzheimer-related changes. Acta Neuropathol 82:239–259

Braak H, Del Tredici K (2011) The pathological process underlying Alzheimer's disease in individuals under thirty. Acta Neuropathol 121:171–181

Bugiani O, Murrell JR, Giaccone G, Hasegawa M, Ghigo G, Tabaton M, Morbin M, Primavera A, Carella F, Solaro C, Grisoli M, Savoiardo M, Spillantini MG, Tagliavini F, Goedert M,

Ghetti B (1999) Frontotemporal dementia and corticobasal degeneration in a family with a P301S mutation in tau. J Neuropathol Exp Neurol 58:667–677

Caughey B, Raymond GJ (1991) The scrapie-associated form of PrP is made from a cell surface precursor that is both protease- and phospholipase-sensitive. J Biol Chem 266:18217–18223

Clavaguera F, Bolmont T, Crowther RA, Abramowski D, Frank S, Probst A, Fraser G, Stalder AK, Beibel M, Staufenbiel M, Jucker M, Goedert M, Tolnay M (2009) Transmission and spreading of tauopathy in transgenic mouse brain. Nat Cell Biol 11:909–913

Colby DW, Prusiner SB (2011) Prions. Cold Spring Harb Perspect Biol 3:a006833

Colby DW, Giles K, Legname G, Wille H, Baskakov IV, DeArmond SJ, Prusiner SB (2009) Design and construction of diverse mammalian prion strains. Proc Natl Acad Sci USA 106:20417–20422

Collinge J, Clarke AR (2007) A general model of prion strains and their pathogenicity. Science 318:930–936

Crowther RA, Goedert M (2000) Abnormal tau-containing filaments in neurodegenerative diseases. J Struct Biol 130:271–279

De Calignon A, Polydoro M, Suárez-Calvet M, William C, Adamowicz DH, Kopeikina KJ, Pitstick R, Sahara N, Ashe KH, Carlson GA, Spires-Jones TL, Hyman BT (2012) Propagation of tau pathology in a model of early Alzheimer's disease. Neuron 73:685–697

Dickinson AG, Meikle VM (1969) A comparison of some biological characteristics of the mouse-passaged scrapie agents, 22A and ME7. Genet Res 13:213–225

Friedhoff P, von Bergen M, Mandelkow EM, Davies P, Mandelkow E (1998) A nucleated assembly mechanism of Alzheimer paired helical filaments. Proc Natl Acad Sci USA 95:15712–15717

Frost B, Jacks RL, Diamond MI (2009a) Propagation of tau misfolding from the outside to the inside of a cell. J Biol Chem 284:12845–12852

Frost B, Ollesch J, Wille H, Diamond MI (2009b) Conformational diversity of wild-type Tau fibrils specified by templated conformation change. J Biol Chem 284:3546–3551

Goedert M, Spillantini MG (2006) A century of Alzheimer's disease. Science 314:777–781

Goedert M, Spillantini MG, Jakes R, Rutherford D, Crowther RA (1989) Multiple isoforms of human microtubule-associated protein tau: sequences and localization in neurofibrillary tangles of Alzheimer's disease. Neuron 3:519–526

Goedert M, Clavaguera F, Tolnay M (2010) The propagation of prion-like protein inclusions in neurodegenerative diseases. Trends Neurosci 33:317–325

Guo JL, Lee VM (2011) Seeding of normal Tau by pathological Tau conformers drives pathogenesis of Alzheimer-like tangles. J Biol Chem 286:15317–15331

Hutton M, Lendon CL, Rizzu P, Baker M, Froelich S, Houlden H, Pickering-Brown S, Chakraverty S, Isaacs A, Grover A, Hackett J, Adamson J, Lincoln S, Dickson D, Davies P, Petersen RC, Stevens M, de Graaff E, Wauters E, van Baren J, Hillebrand M, Joosse M, Kwon JM, Nowotny P, Che LK, Norton J, Morris JC, Reed LA, Trojanowski J, Basun H, Lannfelt L, Neystat M, Fahn S, Dark F, Tannenberg T, Dodd PR, Hayward N, Kwok JB, Schofield PR, Andreadis A, Snowden J, Craufurd D, Neary D, Owen F, Oostra BA, Hardy J, Goate A, van Swieten J, Mann D, Lynch T, Heutink P (1998) Association of missense and 5'-splice-site mutations in tau with the inherited dementia FTDP-17. Nature 393:702–705

Lee VM, Goedert M, Trojanowski JQ (2001) Neurodegenerative tauopathies. Annu Rev Neurosci 24:1121–1159

Liu L, Drouet V, Wu JW, Witter MP, Small SA, Clelland C, Duff K (2012) Trans-synaptic spread of tau pathology in vivo. PLoS One 7:e31302

Nonaka T, Watanabe ST, Iwatsubo T, Hasegawa M (2010) Seeded aggregation and toxicity of {alpha}-synuclein and tau: cellular models of neurodegenerative diseases. J Biol Chem 285:34885–34898

Poorkaj P, Bird TD, Wijsman E, Nemens E, Garruto RM, Anderson L, Andreadis A, Wiederholt WC, Raskind M, Schellenberg GD (1998) Tau is a candidate gene for chromosome 17 frontotemporal dementia. Ann Neurol 43:815–825

Probst A, Götz J, Wiederhold KH, Tolnay M, Mistl C, Jaton AL, Hong M, Ishihara T, Lee VM, Trojanowski JQ, Jakes R, Crowther RA, Spillantini MG, Bürki K, Goedert M (2000) Axonopathy and amyotrophy in mice transgenic for human four-repeat tau protein. Acta Neuropathol 99:469–481

Prusiner SB (1982) Novel proteinaceous infectious particles cause scrapie. Science 216:136–144

Saito Y, Ruberu NN, Sawabe M, Arai T, Tanaka N, Kakuta Y, Yamanouchi H, Murayama S (2004) Staging of argyrophilic grains: an age-associated tauopathy. J Neuropathol Exp Neurol 63:911–918

Spillantini MG, Murrell JR, Goedert M, Farlow MR, Klug A, Ghetti B (1998) Mutation in the tau gene in familial multiple system tauopathy with presenile dementia. Proc Natl Acad Sci USA 95:7737–7741

Tanaka M, Collins SR, Toyama BH, Weissman JS (2006) The physical basis of how prion conformations determine strain phenotypes. Nature 442:585–589

Accumulating Evidence Suggests that Parkinson's Disease Is a Prion-Like Disorder

Nolwen L. Rey, Elodie Angot, Christopher Dunning, Jennifer A. Steiner, and Patrik Brundin

Abstract Parkinson's disease is the second most prevalent neurodegenerative disease, affecting hundreds of thousands of people per year. Although the etiology of the disorder is unclear, recent studies have shown the key involvement of alpha-synuclein in PD. Lewy bodies and Lewy neurites, hallmarks of the disease, contain mostly aggregated alpha-synuclein and develop progressively in the brain in a predictable spatial pattern. It has been proposed that a misfolded form of alpha-synuclein transfers between neighboring cells in a prion-like fashion, acts like a seed, and promotes protein aggregation in those cells, contributing to the progression of the pathology from one brain region to another. Here, we review first the known mechanisms of alpha-synuclein cell-to-cell transfer. We will then discuss the nature of the seed of the aggregation process and, finally, what is missing in the literature to understand the spread of the pathology between brain regions.

Introduction

A major feature of Parkinson's disease (PD) neuropathology is the loss of dopaminergic neurons in the substantia nigra (SN), leading to a dramatic depletion of dopamine in the striatum (Davie 2008). This loss of dopaminergic neurotransmission is believed to be the main cause of motor deficits in PD (Obeso et al. 2008).

Jennifer A. Steiner and Patrik Brundin are co-last authors.

N.L. Rey (✉) • E. Angot • C. Dunning
Neuronal Survival Unit, Wallenberg Neuroscience Center, Lund University, BMC B11,
221 84 Lund, Sweden
e-mail: Nolwen.Rey@med.lu.se

J.A. Steiner
Center for Neurodegenerative Science, Van Andel Research Institute, 333 Bostwick Avenue NE,
Grand Rapids, MI 49503, USA

P. Brundin (✉)
Neuronal Survival Unit, Wallenberg Neuroscience Center, Lund University, BMC B11,
221 84 Lund, Sweden

Center for Neurodegenerative Science, Van Andel Research Institute, 333 Bostwick Avenue NE,
Grand Rapids, MI 49503, USA
e-mail: Patrik.Brundin@med.lu.se

M. Jucker and Y. Christen (eds.), *Proteopathic Seeds and Neurodegenerative Diseases*,
Research and Perspectives in Alzheimer's Disease,
DOI 10.1007/978-3-642-35491-5_8, © Springer-Verlag Berlin Heidelberg 2013

In addition to the loss of dopaminergic neurons, PD is characterized by another histopathological hallmark in the brain: accumulating proteic aggregates rich in alpha-synuclein (α-syn). α-Syn is a 14 kDa protein that is predominantly expressed presynaptically and appears to be involved in vesicle trafficking (Cooper et al. 2006) and in the regulation of the synaptic vesicle pool (Murphy et al. 2000; Nemani et al. 2010). Polymeropoulos et al. (1997) discovered the involvement of α-syn in rare cases of familial autosomal-dominant PD due to a mutation in the SNCA gene, which codes for α-syn protein, opening new perspectives in PD research. We know now that other familial forms of PD, all of which are relatively rare, are also associated with duplications, triplications or mutations of the SNCA gene (Breydo et al. 2012) or with mutations of genes coding for parkin (PRKN), leucin-rich repeat kinase 2 (LRRK2), PTEN-induced putative kinase 1 (PINK1), VPS35, and DJ1 (Davie 2008; Vilarino-Guell et al. 2011; Zimprich et al. 2011).

Shortly after the discovery of the first mutations in the SNCA gene, it was revealed that the abnormal cytoplasmic protein inclusions in PD contain fibrillar aggregates of α-syn (Spillantini et al. 1997; Goedert 2001). The α-syn in the aggregates is phosphorylated (Fujiwara et al. 2002; Neumann et al. 2002) and the aggregates also contain ubiquitin (Gai et al. 2000). They are found in the cell bodies of neurons and in neurites, where they are called Lewy bodies (LB) or Lewy neurites (LN), respectively. Occasionally, aggregated α-syn is also found in glia, e.g., in oligodendroglia in multiple system atrophy (Wakabayashi and Takahashi 2006).

Detailed studies of LN and LB in PD by Braak and colleagues (Braak et al. 2003a, b) have suggested that α-syn aggregate pathology progresses from certain brain regions to others following a predictable pattern of six stages. Specifically, Braak and colleagues hypothesized that a neurotropic pathogen entered the brain after ingestion or inhalation and that the point(s) of entry might be olfactory structures and the gut. Once the pathogen had gained access to the olfactory pathways and the enteric nervous system, they proposed that it transferred between cells to eventually reach the SN after several years (Hawkes et al. 2007), causing dysfunction and death of dopaminergic neurons and the appearance of motor symptoms. We put forward the hypothesis that α-syn might itself initially be targeted by such a pathogen, leading to its misfolding, and that misfolded α-syn thus plays the role of the spreading agent when the neuropathological changes propagate to other brain regions (Steiner et al. 2011). Indeed, recent studies have focused on the possibility of a seeding mechanism and a transfer from one cell to another of misfolded α-syn in a prion-like fashion in PD and how this might explain the observations of Braak and colleagues.

In this chapter, we first review what is known regarding mechanisms of α-syn misfolding and the cellular mechanisms of cell-to-cell transfer. Further, we discuss the identity of the transmissible seed of the α-syn aggregation process and the molecular nature of the seed (monomers, oligomers and fibrils). Finally, we consider what parts of the puzzle are currently missing for a better understanding of how synucleinopathy spreads between brain regions.

PD Progression and α-Syn Cell-to-Cell Transfer: Short Review of Our Current Knowledge

In prion diseases, prion protein (PrP^C) can adopt incorrect folding and become prone to aggregation. Misfolded prion protein (PrP^{Sc}) acts as a nucleus that recruits other proteins, a process called "nucleation," which leads to the formation of aggregates that become infectious (Prusiner 1982, 1998). This process can lead to prion diseases, such as Creutzfeldt-Jakob Disease, that typically are devastating and rapidly progressing once the first symptoms have appeared.

However, what relationship might prion diseases have with slowly progressing neurodegenerative diseases such as PD? A mechanism similar to that of prion disease was proposed for PD as early as 7 years ago (Prusiner 2001; Hardy 2005). Both high expression levels of α-syn and mutations of the SNCA gene promote the deposition of misfolded α-syn. Therefore, Hardy (2005) suggested that, depending on the concentration of the pathogenic protein, a rare event might initiate the disease by creating a pathogenic template.

This hypothesis of a permissive templating in PD was brought to center stage following unexpected and dramatic results from anatomical studies of people with PD who had received neuronal grafts of embryonic mesencephalic tissues into the striatum over 10 years prior to their deaths (Kordower et al. 2008; Li et al. 2008, 2010). In these patients, LB and LN developed in a small percentage of young grafted neurons. Featuring misfolded α-syn, these inclusions bore all the hallmarks of LB identified in aged host neurons (Li et al. 2010).

Based on these and other findings, we proposed that a misfolded α-syn protein derived from a host cell might penetrate and induce aggregation of endogenous α-syn in a prion-like manner in neighboring grafted neurons (Brundin et al. 2008; Li et al. 2008). Our recent studies and those of others, all performed in rodents, bolstered the evidence that a cell-to-cell transfer of α-syn could occur in vivo (Desplats et al. 2009; Hansen et al. 2011; Kordower et al. 2011; Mougenot et al. 2011; Angot et al. 2012). Companion in vitro studies have provided many clues to elucidate the molecular mechanisms that underlie such a spread. We hypothesize that this spread of α-syn occurs in a three-step process, starting with the secretion of misfolded α-syn from a donor cell, followed by the uptake of this template by a recipient cell, and culminating in α-syn acting like a seed and inducing nucleation in the recipient cell (reviewed in Angot and Brundin 2009; Angot et al. 2010; Dunning et al. 2011; Steiner et al. 2011).

α-Syn Misfolding and Aggregation

The first step of a pathogenetic process involving a prion-like spread of α-syn is the misfolding of the protein. In a pathogenic environment such as the parkinsonian brain, natively unfolded α-syn is somehow transformed into a beta-sheet-rich protein (the misfolded form) that assembles into oligomers and aggregates into amyloid fibrils, which are the basis of LN and LB.

Natively unfolded α-syn can adopt different molecular states (monomers, oligomers and fibrils) or different conformations (for details, see below), depending on changes in the environment and interaction with cofactors and lipids (Bisaglia et al. 2009; Auluck et al. 2010; Oueslati et al. 2010; Papy-Garcia et al. 2011; Stefanis 2012). α-Syn can also be modified by various post-translational modifications, such as phosphorylation, nitrosylation, or glycosylation.

The fact that α-syn was found in aggregated and phosphorylated forms in LB suggested that changes in protein conformation and post-translational modifications, e.g., α-syn phosphorylation in LB (Fujiwara et al. 2002; Neumann et al. 2002; Stefanis 2012), might be essential steps in the pathogenetic process. However, recent studies have cast doubt on the toxicity of phosphorylated α-syn and its capacity to induce aggregation (Gorbatyuk et al. 2008; Paleologou et al. 2008; Azeredo da Silveira et al. 2009). While it is still unclear what causes α-syn to misfold, many in vitro studies have been performed to understand the process of α-syn aggregation (for extensive review, see Breydo et al. 2012).

Wood et al. (1999) demonstrated that misfolded α-syn can induce aggregation of natively unfolded α-syn, Based on these studies and others, it appears that soluble monomeric α-syn binds to the misfolded α-syn seed reversibly and further undergoes an irreversible reorganization that produces a more stable aggregate, in a fibrillar organization. By repetitions of these mechanisms, fibrils extend (Breydo et al. 2012).

Although the steps of nucleation and extension are relatively well understood, it is not clear at which point in the process cell-to-cell transfer occurs. Once misfolded (and perhaps aggregated), α-syn can occasionally transfer to another cell, as shown by the in vivo and in vitro studies we describe in the following sections (Desplats et al. 2009; Hansen et al. 2011; Kordower et al. 2011).

Evidence for Cell-to-Cell Transfer and Possible Mechanisms

Recent In Vivo Studies

Our investigations and those of other groups provide strong support for the hypothesis that α-syn can transfer from one cell to another and can contribute to the spreading of Lewy pathology. To mimic the grafting of mesencephalic neurons into the striatum of PD patients (Kordower et al. 2008; Li et al. 2008, 2010; Mendez et al. 2008), we grafted ventral embryonic mouse mesencephalon into the striatum of adult transgenic mice overexpressing human α-syn. Six months after grafting, we found human α-syn puncta inside select tyrosine hydroxylase (TH)-positive cells of the graft (Hansen et al. 2011). These results demonstrate that human α-syn can transfer from the host striatum to grafted dopaminergic neurons and support the notion that this process also had occurred in the brains of the PD patients displaying LB inside grafted neurons.

The host-to-graft pathology transfer process suspected in transplanted PD patients has also been reproduced in rats with 6-hydroxydopamine lesions of the nigrostriatal pathway. These rats, which lacked dopaminergic input to the striatum akin to what is seen in PD, were then given intrastriatal grafts of fetal rat midbrain tissue. One month after the graft, the investigators used injections of viral vectors to overexpress human α-syn [or Green Fluorescent Protein (GFP) as a control] in the striatum rostral to the graft (Kordower et al. 2011). Five weeks after virus injection, some grafted dopaminergic neurons that were not in immediate, apparent contact with the virus injection site displayed host-derived α-syn but no GFP, indicating a specific transfer of α-syn from the host to grafted cells.

Interestingly, Desplats et al. (2009) also showed that, in a transgenic mouse model overexpressing α-syn, α-syn could transfer from host cells to neuronal stem cells grafted in the hippocampus. This result indicates that the host-to-graft transfer is not specific to grafts in the striatum.

Recent In Vitro Studies

Release of α-Syn by Cells

The mechanism of α-syn cell-to-cell transfer is not fully understood, but recent in vitro investigations have shed some light on the different steps involved.

The first step of a cell-to-cell transfer is the exit of α-syn from the host cell. Living cells can release α-syn into the extracellular space through exocytosis in vitro, a process independent of the classic endoplasmic reticulum/Golgi pathway for secretion (Lee et al. 2005). And interestingly, the release of α-syn is enhanced when cells are exposed to stress conditions (Jang et al. 2010), suggesting that unhealthy, and maybe dying, cells might also release high quantities of misfolded α-syn, resulting in raised extracellular levels of α-syn. Interestingly, α-syn can be released locally from axonal terminals in vitro (Danzer et al. 2011), suggesting that a cell-to-cell transfer might occur trans-synaptically.

In addition to classical exocytosis, recent studies suggest a possible role of exosomes in the cellular release of α-syn. Exosomes are 40–100 nm vesicles of endocytic origin that are involved in the removal of unwanted proteins from cells and in cell-to-cell communication (Simons and Raposo 2009). Recently, it has been shown that exosomes are involved in α-syn secretion from cultured SH-SY5Y cells and that this release is enhanced when the cell undergoes lysosomal dysfunction (Emmanouilidou et al. 2010; Alvarez-Erviti et al. 2011).

In conclusion, two different mechanisms have been proposed for α-syn release by a cell, and, in both cases, cellular dysfunction such as that which might occur in early PD could enhance the amount of released α-syn.

Uptake of α-Syn by Recipient Cells and Nucleation

The next step in the disease process, following the release of α-syn by the donor cell, is the entrance of the protein into a recipient cell. The way this event takes place clearly depends on which form of α-syn is being imported. Indeed, monomeric α-syn, with its low molecular weight, can passively and very quickly (within 5 min) translocate into cultured recipient cells through the outer plasma membrane (Ahn et al. 2006). By contrast, oligomers (Danzer et al. 2009) and fibrils formed using recombinant human α-syn are taken up more slowly and often require the use of lipid based reagents to facilitate entry (Lee et al. 2008a, b). This slow uptake has been shown to occur via a classic endocytotic mechanism (Lee et al. 2008a; Hansen et al. 2011); however, the exact mechanisms and whether or not receptors are involved in this process remain unclear (Ahn et al. 2006).

We have also demonstrated in vivo that recombinant human α-syn injected into the neocortex of rodents was taken up after 6 h by cortical neurons and this uptake was reduced upon co-injection with dynasore, an inhibitor of endocytosis (Hansen et al. 2011). These data showed that α-syn can be internalized in neurons at least in part by endocytosis in vivo as well as in vitro (Hansen et al. 2011).

For the prion-like hypothesis to be valid, it is not sufficient that misfolded α-syn be taken up by recipient cells; it must also attract endogenous α-syn and induce its misfolding. The possibility that this process might occur was illustrated when cultured primary cortical neurons not only took up specific forms of α-syn oligomers but also started to exhibit punctate α-syn staining suggestive of aggregation, as opposed to a homogeneous distribution that would have been consistent with the soluble form of the protein (Danzer et al. 2007, 2009).

Recently, we obtained further evidence of nucleation occurring after α-syn cell-to-cell transfer. We used a co-culture model where cells expressed either GFP-α-syn or DsRed-α-syn (Hansen et al. 2011). We found numerous double-labeled cells, indicating a transfer of α-syn between cells. Importantly, confocal microscope images revealed that, in some cells, GFP-α-syn was enveloped by endogenous DsRed-α-syn. This finding strongly suggested that a seeding mechanism induced by the transferred protein, leading to accumulation of endogenous protein around imported proteins, could occur in the recipient cell (Hansen et al. 2011).

A recruitment of endogenous α-syn has also been documented in the recent work of Volpicelli-Daley et al. (2011), where the authors showed that preformed fibrils (derived from full-length and truncated α-syn) could enter primary neurons in cell culture, recruit endogenous α-syn, and form α-syn clusters inside the recipient cell. The authors then went on to assess the consequences of α-syn uptake into the recipient cells. Interestingly, this transfer of α-syn into recipient cells led to a reduction of synaptic protein expression, progressive alterations of neuronal excitability and coordination in neuronal assembly, and eventually caused cell death (Volpicelli-Daley et al. 2011).

Besides classic endocytic mechanisms, two other mechanisms have been proposed for the entrance of α-syn in a recipient cell: exosomes and tunneling nanotubes (TNTs).

First, Alvarez-Erviti et al. (2011) observed that exosomes containing α-syn were able occasionally to transfer the protein to new cells, forming inclusions into these cells. However, the nature of those clusters of protein is unclear. The observed α-syn could be either vesicle bound or aggregated protein.

Secondly, TNTs, which are communication systems that mediate long-range communication through long processes, are known to allow the exchange of small molecules, organelles and vesicles between mammalian cells (Gerdes et al. 2007). Recently, α-syn has been reported to exist inside TNTs, and it has been speculated that it might migrate through that route from one cell to another (Agnati et al. 2010). Even if the uptake of α-syn by classic endocytosis has been demonstrated, other mechanisms such as exosome- and TNT-mediated transfer of α-syn need to be further investigated, as cells under stress may use alternate or additional methods of transfer.

Intracellular Transport of α-Syn

We assume at this point that misfolded α-syn can transfer from a cell to neighboring cells. But how could the pathology propagate to other regions of the brain, based on Braak's observations in people with PD?

The investigations of Braak and colleagues suggested that LN and LB appear first in the anterior olfactory structures and the vagal nerve and then spread to interconnected regions such as the amygdala and the brainstem, and then to neocortex (Braak et al. 2003a, b, 2004). Based on their findings, they proposed the idea that α-syn pathology spread follows long and unmyelinated axons and that the unknown pathogenic agent could be transported retrogradely or anterogradely in axons (Braak et al. 2003a; Hawkes et al. 2007, 2009).

Volpicelli-Daley and colleagues have demonstrated the real-time transport of imported α-syn by in vitro methods. They used microfluidic cultures in which neurites could be isolated from cell bodies and added preformed α-syn fibrils either at the level of the cell body or near neurites. With this approach, they showed for the first time that imported α-syn could be transported intracellularly, both retrogradely and anterogradely in vitro (Volpicelli-Daley et al. 2011). Recently, the same laboratory demonstrated that α-syn fibrils can spread along axonal pathways in the living brains of transgenic mice expressing human A53T mutant α-syn. They stereotaxically injected preformed α-syn fibrils in the cortex and striatum and detected hyperphosphorylated and misfolded α-syn at sites distant from the injection sites 90 days later. The most severe pathology developed in the most densely interconnected areas. As a consequence of the widespread synucleinopathy, the mice died earlier than uninjected transgenic mice (Luk et al. 2012).

These two initial studies demonstrating intercellular transport of α-syn represent important proofs-of-principle. They show that the phenomenon warrants further investigation and should stimulate experiments exploring factors that affect intraneuronal transport of misfolded α-syn.

Despite the recent research that has shed light on possible mechanisms accounting for cell-to-cell transfer of pathological α-syn and its transmission (for an extended review of cell-to-cell transfer with regard to synucleinopathies, see Angot and Brundin 2009; Angot et al. 2010; Dunning et al. 2011; Steiner et al. 2011), many questions still remain. In the next section, we will discuss the possible origin and nature of the seed and review what evidence is still needed to prove that PD is also a prion-like disease.

Origin of the Seeding and Transfer In Vivo: What Is Missing in the Literature?

What Causes α-Syn Seeding?

It is assumed that, for α-syn to recruit and seed aggregation, the donor α-syn itself must be corrupted in some manner. This could be the case of the monomer being misfolded in a situation similar to prion toxicity, or the more likely scenario where misfolding of monomeric α-syn results in the formation of toxic oligomers and/or fibrils. With this in mind, perhaps the most pressing question is, "What is the native conformation of α-syn?" Previously α-syn was thought to be a monomeric protein with little to no secondary structure. Recently, research from two groups suggested that endogenous and bacterially expressed α-syn exists as a stable tetramer (Bartels et al. 2011; Wang et al. 2011). This oligomeric species of α-syn is proposed to be resistant to aggregation and, as such, is most likely to represent the functional form of the protein. Attempts to recapitulate these experiments have proved unsuccessful, while still others have questioned the techniques utilized by the groups mentioned above (Fauvet et al. 2012) (Debated in Azheimer Research Forum, http://www.alzforum.org/pap/annotation.asp?powID=128115). Some agreement has been reached that α-syn is likely to exist in an equilibrium, possibly based on function and/or location, between monomer and oligomer. However, it remains hotly debated which α-syn state is more prevalent and functionally and pathologically relevant,

The question of the choice of protocol is also crucial to the question of how aggregation occurs and what form(s) of α-syn favor this process. In cell culture models of α-syn spread, protein aggregation is a rare event. Perhaps this is because aggregation is more likely to occur in differentiated cells than in the proliferating cells that a number of groups, ourselves included, use (Hansen et al. 2011). Observing aggregation using recombinant protein, however, is much more achievable (Danzer et al. 2009; Volpicelli-Daley et al. 2011). Despite this reality, inconsistencies exist in the literature regarding the species of recombinant α-syn used. It is not as simple as having each laboratory denote their α-syn preparations as monomeric, oligomeric or fibrillar, because they typically each use unique protocols that potentially introduce differences. For instance, research by Dr. Karin Danzer et al. (2009, 2011) revealed that oligomers of α-syn prepared using different protocols resulted in altered uptake, seeding and toxicity. Recombinant fibrils have also been shown to cause seeding, however; reports out of the laboratory of Dr. Virginia Lee showed that one protocol for fibril-induced seeding required lipid based transfection reagents for cellular entry (Luk et al. 2009), whereas others suggest recombinant fibrils (Volpicelli-Daley et al. 2011; Luk et al. 2012) and α-syn from brain homogenates (Luk et al. 2012) can be taken up directly by the cell. A similar debate has been ongoing in the AD field, with researchers seeking consistency on the type of Aβ fibril used for studies (Editorial, Nature Neuroscience 2011). An important question is which species best represents the pathogenic species occurring in the brain and what is the best method of production and experimental design to recapitulate this.

Finally, we need also to understand why pathology progressively involves greater numbers of brain regions. Are particular neurons more susceptible to PD pathology, i.e., less apt to dispose of misfolded α-syn that has been taken up from the extracellular space, than others? Does the pathology only spread from a cell to neighboring cells, as the Braak hypothesis would suggest, or, perhaps less likely, can α-syn move over larger distances in the extracellular space? Certainly direct transfer between neighboring cells makes sense, considering synaptic connectivity and the temporal progression of the lesions in the CNS; however, the presence of α-syn in exosomes and the extracellular space [including human cerebrospinal fluid (CSF)] seems to suggest that cell-to-cell contact is not necessary. Why then do we not see pathology spreading in a random manner among a whole range of different neurons? One possible explanation is that misfolded α-syn may hijack exosomes that are targeted to specific cells for signaling purposes, or it could be that cells take up exosomes in an indiscriminate manner and that certain cells are better equipped to deal with the potentially deadly cargo carried by the exosomes. It is curious to note that PD is characterized by the death of dopaminergic neurons in the SN, whereas in the neighboring ventral tegmental area (VTA), dopaminergic neurons are largely unaffected until later stages in the disease. So does the pathology spread from the SN to the VTA? Perhaps it is the different energy requirements of the neurons that make those in the SN more susceptible (Moss and Bolam 2008). This energy requirement could also be linked to calcium and the autonomous pacemaking activity of neurons in the SN (Guzman et al. 2009; Surmeier et al. 2011). Interestingly, neurons with a similar pacemaking activity are found in the locus coeruleus (Williams et al. 1984), dorsal motor nucleus of the vagus nerve (DMV) (McCann and Rogers 1990) and olfactory bulb (Pignatelli et al. 2005). All of these regions are outlined by Braak and colleagues in their staging hypothesis for PD pathology (Braak et al. 2003a, b, 2004).

In Vivo Evidence for Prion-Like Propagation of α-Syn: What Is Missing in the Literature?

Can α-Syn Induce a Nucleation Process After Cell-to-Cell Transfer?

As described above, all the steps of α-syn prion-like propagation—release from donor cell, uptake into recipient cell, seeding of endogenous α-syn proteins from the recipient cell around a core of transferred α-syn—have been observed in in vitro models. The precise molecular mechanisms, however, remain to be elucidated. In experimental animals, transfer of α-syn was shown from host brain to grafted neural stem cells (Desplats et al. 2009) and embryonic dopaminergic neurons (Hansen et al. 2011). Moreover, brain extracts from old transgenic mice containing insoluble and phosphorylated forms of α-syn accelerated the synucleinopathy in young, transgenic mice expressing human A53T mutant α-syn (Mougenot et al. 2011). In their report, Mougenot et al. (2011) presented the first proof of a seeding effect of α-syn in vivo. In these experiments, the "seed" was exogenously presented to the extracellular space in the brain by intracerebral inoculation. What remains to be reported is a nucleation effect of α-syn after transfer from one neuron to another. We have recently developed a rat model in which this phenomenon does indeed occur. We transplanted rat embryonic ventral mesencephalon in the striatum of rats overexpressing human α-syn in the nigrostriatal pathway as a consequence of an intranigral injection of Adeno-Associated Vectors (AAV) carrying the human α-syn gene. We studied the effect of varying the intervals between viral transduction and graft surgery, as well as grafting and sacrifice, on the frequency of cells exhibiting α-syn transfer. Under optimal conditions, we found that 23 % of grafted rat dopaminergic neurons displayed intracellular human α-syn immunoreactivity (Angot et al. 2012). In rare instances, we observed a core of human α-syn embedded in a shell of rat α-syn within the transplanted TH-expressing neurons, suggesting a seeding effect of the transferred human α-syn on the endogenous α-syn proteins expressed by the recipient cell (Angot et al. 2012).

Can α-Syn Transfer from One Brain Region to Another?

The prion-like spread of α-syn could account for the propagation of α-synucleinopathy between interconnected brain regions proposed by Braak et al., and as mentioned above this propagation has now been modeled in rodents following intracerebral injection of α-syn preformed fibrils (Luk et al. 2012). Two independent research groups recently reported a similar model for tauopathy spread during the early stages of AD (de Calignon et al. 2012; Liu et al. 2012) Both groups developed a transgenic mouse line expressing pathogenic forms of tau exclusively in entorhinal cortex, the first structure displaying tau tangles in early Alzheimer's disease (AD; Braak and Braak 1991). In young mice, tau immunoreactivity was indeed restricted to the entorhinal cortex, whereas in older mice, tau deposits were found in other brain areas synaptically connected to entorhinal cortex but devoid of any detectable transgene expression, notably the dentate gyrus and different fields of hippocampal pyramidal neurons (de Calignon et al. 2012; Liu et al. 2012).

Based on the progression of LB and LN in PD, Hawkes et al. hypothesized that PD starts with a dual-hit, comprising a simultaneous entry of a pathogen into the brain through a nasal route and a route involving the enteric nerves that innervate the gastrointestinal system (Hawkes et al. 2007, 2009). To test this hypothesis in mice, one could restrict the expression of abnormal α-syn proteins to the gut and/or the olfactory bulb and investigate whether α-syn-related pathology propagates to interconnected brain areas.

Can Different Conformations of α-Syn Induce Different Patterns of Synucleinopathies?

In mouse models of AD, the pattern of amyloidosis induced in the brain of transgenic hosts depends on the nature of the Aβ in the brain extracts injected intracerebrally (Meyer-Luehmann et al. 2006). Whether different conformations or "strains" of α-syn induce different patterns of synuclein-related pathology remains to be addressed systematically. Differences in α-syn "strains" that initially trigger the disease might account for the differences in rates of disease progression and which neuronal populations are affected in the different synucleinopathies that occur in humans. Whether synucleinopathy can be induced by exposure to pathogenic forms of α-syn in the systemic blood circulation is worth examining, not least because α-syn is enriched in red blood cells (Barbour et al. 2008). Results of such studies might have major clinical implications, as they could implicate blood transfusion as a risk factor for PD.

Can α-Syn Cross-seed with Other Amyloid Proteins?

Another possible explanation for a primary pathogenic event leading to α-syn misfolding is the cross-seeding with other amyloid proteins that naturally occurs in our environment. Silk from the butterfly *Bombyx mori* or curli from the bacteria *Escherichia coli* cross-seed the aggregation of amyloid protein A (Lundmark et al. 2005), which deposits in different organs of patients with chronic inflammatory disease, as rheumatoid arthritis (Kobayashi et al. 1996). In vitro, prion protein and Aβ aggregation can be enhanced by a cross-seeding mechanism (Morales et al. 2010). Moreover, inoculation of transgenic mouse models of AD with prions accelerates and exacerbates both pathologies (Morales et al. 2010). No such evidence for cross-seeding has been reported so far for α-syn, but α-syn inclusions and Aβ deposits do occur in the brains of patients with PD and Dementia with Lewy Bodies (DLB; Goedert 2001; Norris et al. 2004). In addition, patients with familial AD develop cerebral α-syn aggregation (Lippa et al. 1998; Ishikawa et al. 2005).

However, in all these cases, the two protein deposits are not located in the same inclusions, which might suggest that any interaction occurring between these two misfolded proteins is indirect.

Conclusions

Even though recent studies have provided strong support and offered up possible mechanisms to explain the prion-like spread of a-syn, this field is still in its infancy and many questions remain unanswered. Nevertheless, the recent studies open up new avenues for PD treatment. Any molecule able to interfere with one of the four steps of this process (misfolding, release from the donor cell, uptake by the recipient cell and seeding of aggregation of endogenous α-syn proteins) could potentially slow down or even stop the progression of synucleinopathy (Brundin and Olsson 2011).

With the prion-like hypothesis applying to the spread of other neurodegenerative disease besides PD, perhaps some answers lie in research already undertaken outside the PD field. What is clear, however, is that the prion-like hypothesis is becoming more accepted and is likely to be a major research focus at the pre-clinical, diagnostic and therapeutic level for years to come.

Acknowledgments Our work discussed above was supported by a European Research Council Advanced Grant and by the Swedish Research Council, Human Frontier Science Program, Parkinson Foundation in Sweden, ERA-Net NEURON–MIPROTRAN, and by the BAGADILICO Linnaeus environment and MultiPark strategic research area at Lund University, both of which are sponsored by the Swedish Research Council. These funding agencies did not influence the content of this review.

References

Agnati LF, Guidolin D, Baluska F, Leo G, Barlow PW, Carone C, Genedani S (2010) A new hypothesis of pathogenesis based on the divorce between mitochondria and their host cells: possible relevance for Alzheimer's disease. Curr Alzheimer Res 7:307–322

Ahn KJ, Paik SR, Chung KC, Kim J (2006) Amino acid sequence motifs and mechanistic features of the membrane translocation of alpha-synuclein. J Neurochem 97:265–279

Alvarez-Erviti L, Seow Y, Schapira AH, Gardiner C, Sargent IL, Wood MJ, Cooper JM (2011) Lysosomal dysfunction increases exosome-mediated alpha-synuclein release and transmission. Neurobiol Dis 42:360–367

Angot E, Brundin P (2009) Dissecting the potential molecular mechanisms underlying alpha-synuclein cell-to-cell transfer in Parkinson's disease. Parkinsonism Relat Disord 15(Suppl 3): S143–S147

Angot E, Steiner JA, Hansen C, Li JY, Brundin P (2010) Are synucleinopathies prion-like disorders? Lancet Neurol 9:1128–1138

Angot E, Steiner JA, Lema Tomé CM, Ekström P, Mattsson B, Björklund A, Brundin P (2012) Alpha-synuclein cell-to-cell transfer and seeding in grafted dopaminergic neurons in vivo. PLoS One 7(6):e39465

Auluck PK, Caraveo G, Lindquist S (2010) alpha-Synuclein: membrane interactions and toxicity in Parkinson's disease. Annu Rev Cell Dev Biol 26:211–233

Azeredo da Silveira S, Schneider BL, Cifuentes-Diaz C, Sage D, Abbas-Terki T, Iwatsubo T, Unser M, Aebischer P (2009) Phosphorylation does not prompt, nor prevent, the formation of alpha-synuclein toxic species in a rat model of Parkinson's disease. Hum Mol Genet 18:872–887

Barbour R, Kling K, Anderson JP, Banducci K, Cole T, Diep L, Fox M, Goldstein JM, Soriano F, Seubert P, Chilcote TJ (2008) Red blood cells are the major source of alpha-synuclein in blood. Neurodegener Dis 5:55–59

Bartels T, Choi JG, Selkoe DJ (2011) alpha-Synuclein occurs physiologically as a helically folded tetramer that resists aggregation. Nature 477:107–110

Bisaglia M, Tessari I, Mammi S, Bubacco L (2009) Interaction between alpha-synuclein and metal ions, still looking for a role in the pathogenesis of Parkinson's disease. Neuromolecular Med 11:239–251

Braak H, Braak E (1991) Neuropathological stageing of Alzheimer-related changes. Acta Neuropathol 82:239–259

Braak H, Del Tredici K, Rub U, de Vos RA, Jansen Steur EN, Braak E (2003a) Staging of brain pathology related to sporadic Parkinson's disease. Neurobiol Aging 24:197–211

Braak H, Rub U, Gai WP, Del Tredici K (2003b) Idiopathic Parkinson's disease: possible routes by which vulnerable neuronal types may be subject to neuroinvasion by an unknown pathogen. J Neural Transm 110:517–536

Braak H, Ghebremedhin E, Rub U, Bratzke H, Del Tredici K (2004) Stages in the development of Parkinson's disease-related pathology. Cell Tissue Res 318:121–134

Breydo L, Wu JW, Uversky VN (2012) Alpha-synuclein misfolding and Parkinson's disease. Biochim Biophys Acta 1822:261–285

Brundin P, Olsson R (2011) Can alpha-synuclein be targeted in novel therapies for Parkinson's disease? Expert Rev Neurother 11:917–919

Brundin P, Li JY, Holton JL, Lindvall O, Revesz T (2008) Research in motion: the enigma of Parkinson's disease pathology spread. Nat Rev Neurosci 9:741–745

Cooper AA, Gitler AD, Cashikar A, Haynes CM, Hill KJ, Bhullar B, Liu K, Xu K, Strathearn KE, Liu F, Cao S, Caldwell KA, Caldwell GA, Marsischky G, Kolodner RD, Labaer J, Rochet JC, Bonini NM, Lindquist S (2006) Alpha-synuclein blocks ER-Golgi traffic and Rab1 rescues neuron loss in Parkinson's models. Science 313:324–328

Danzer KM, Haasen D, Karow AR, Moussaud S, Habeck M, Giese A, Kretzschmar H, Hengerer B, Kostka M (2007) Different species of alpha-synuclein oligomers induce calcium influx and seeding. J Neurosci 27:9220–9232

Danzer KM, Krebs SK, Wolff M, Birk G, Hengerer B (2009) Seeding induced by alpha-synuclein oligomers provides evidence for spreading of alpha-synuclein pathology. J Neurochem 111:192–203

Danzer KM, Ruf WP, Putcha P, Joyner D, Hashimoto T, Glabe C, Hyman BT, McLean PJ (2011) Heat-shock protein 70 modulates toxic extracellular alpha-synuclein oligomers and rescues trans-synaptic toxicity. FASEB J 25:326–336

Davie CA (2008) A review of Parkinson's disease. Br Med Bull 86:109–127

de Calignon A, Polydoro M, Suarez-Calvet M, William C, Adamowicz DH, Kopeikina KJ, Pitstick R, Sahara N, Ashe KH, Carlson GA, Spires-Jones TL, Hyman BT (2012) Propagation of tau pathology in a model of early Alzheimer's disease. Neuron 73:685–697

Desplats P, Lee HJ, Bae EJ, Patrick C, Rockenstein E, Crews L, Spencer B, Masliah E, Lee SJ (2009) Inclusion formation and neuronal cell death through neuron-to-neuron transmission of alpha-synuclein. Proc Natl Acad Sci USA 106:13010–13015

Dunning CJ, Reyes JF, Steiner JA, Brundin P (2011) Can Parkinson's disease pathology be propagated from one neuron to another? Prog Neurobiol 97:205–219

Editorial (2011) State of aggregation. Nat Neurosci 14:399

Emmanouilidou E, Melachroinou K, Roumeliotis T, Garbis SD, Ntzouni M, Margaritis LH, Stefanis L, Vekrellis K (2010) Cell-produced alpha-synuclein is secreted in a calcium-dependent manner by exosomes and impacts neuronal survival. J Neurosci 30:6838–6851

Fauvet B, Mbefo MK, Fares MB, Desobry C, Michael S, Ardah MT, Tsika E, Coune P, Prudent M, Lion N, Eliezer D, Moore DJ, Schneider B, Aebischer P, El-Agnaf OM, Masliah E, Lashuel HA (2012) Alpha-synuclein in the central nervous system and from erythrocytes, mammalian cells and E. coli exists predominantly as a disordered monomer. J Biol Chem 287 (19):15345–15364

Fujiwara H, Hasegawa M, Dohmae N, Kawashima A, Masliah E, Goldberg MS, Shen J, Takio K, Iwatsubo T (2002) alpha-Synuclein is phosphorylated in synucleinopathy lesions. Nat Cell Biol 4:160–164

Gai WP, Yuan HX, Li XQ, Power JT, Blumbergs PC, Jensen PH (2000) In situ and in vitro study of colocalization and segregation of alpha-synuclein, ubiquitin, and lipids in Lewy bodies. Exp Neurol 166:324–333

Gerdes HH, Bukoreshtliev NV, Barroso JF (2007) Tunneling nanotubes: a new route for the exchange of components between animal cells. FEBS Lett 581:2194–2201

Goedert M (2001) Alpha-synuclein and neurodegenerative diseases. Nat Rev Neurosci 2:492–501

Gorbatyuk OS, Li S, Sullivan LF, Chen W, Kondrikova G, Manfredsson FP, Mandel RJ, Muzyczka N (2008) The phosphorylation state of Ser-129 in human alpha-synuclein determines neurodegeneration in a rat model of Parkinson disease. Proc Natl Acad Sci USA 105:763–768

Guzman JN, Sanchez-Padilla J, Chan CS, Surmeier DJ (2009) Robust pacemaking in substantia nigra dopaminergic neurons. J Neurosci 29:11011–11019

Hansen C, Angot E, Bergstrom AL, Steiner JA, Pieri L, Paul G, Outeiro TF, Melki R, Kallunki P, Fog K, Li JY, Brundin P (2011) alpha-Synuclein propagates from mouse brain to grafted dopaminergic neurons and seeds aggregation in cultured human cells. J Clin Invest 121:715–725

Hardy J (2005) Expression of normal sequence pathogenic proteins for neurodegenerative disease contributes to disease risk: 'permissive templating' as a general mechanism underlying neurodegeneration. Biochem Soc Trans 33:578–581

Hawkes CH, Del Tredici K, Braak H (2007) Parkinson's disease: a dual-hit hypothesis. Neuropathol Appl Neurobiol 33:599–614

Hawkes CH, Del Tredici K, Braak H (2009) Parkinson's disease: the dual hit theory revisited. Ann N Y Acad Sci 1170:615–622

Ishikawa A, Piao YS, Miyashita A, Kuwano R, Onodera O, Ohtake H, Suzuki M, Nishizawa M, Takahashi H (2005) A mutant PSEN1 causes dementia with Lewy bodies and variant Alzheimer's disease. Ann Neurol 57:429–434

Jang A, Lee HJ, Suk JE, Jung JW, Kim KP, Lee SJ (2010) Non-classical exocytosis of alpha-synuclein is sensitive to folding states and promoted under stress conditions. J Neurochem 113:1263–1274

Kobayashi H, Tada S, Fuchigami T, Okuda Y, Takasugi K, Matsumoto T, Iida M, Aoyagi K, Iwashita A, Daimaru Y, Fujishima M (1996) Secondary amyloidosis in patients with rheuma-toid arthritis: diagnostic and prognostic value of gastroduodenal biopsy. Br J Rheumatol 35:44–49

Kordower JH, Chu Y, Hauser RA, Freeman TB, Olanow CW (2008) Lewy body-like pathology in long-term embryonic nigral transplants in Parkinson's disease. Nat Med 14:504–506

Kordower JH, Dodiya HB, Kordower AM, Terpstra B, Paumier K, Madhavan L, Sortwell C, Steece-Collier K, Collier TJ (2011) Transfer of host-derived alpha synuclein to grafted dopaminergic neurons in rat. Neurobiol Dis 43:552–557

Lee HJ, Patel S, Lee SJ (2005) Intravesicular localization and exocytosis of alpha-synuclein and its aggregates. J Neurosci 25:6016–6024

Lee HJ, Suk JE, Bae EJ, Lee JH, Paik SR, Lee SJ (2008a) Assembly-dependent endocytosis and clearance of extracellular alpha-synuclein. Int J Biochem Cell Biol 40:1835–1849

Lee HJ, Suk JE, Bae EJ, Lee SJ (2008b) Clearance and deposition of extracellular alpha-synuclein aggregates in microglia. Biochem Biophys Res Commun 372:423–428

Li JY, Englund E, Holton JL, Soulet D, Hagell P, Lees AJ, Lashley T, Quinn NP, Rehncrona S, Bjorklund A, Widner H, Revesz T, Lindvall O, Brundin P (2008) Lewy bodies in grafted neurons in subjects with Parkinson's disease suggest host-to-graft disease propagation. Nat Med 14:501–503

Li JY, Englund E, Widner H, Rehncrona S, Bjorklund A, Lindvall O, Brundin P (2010) Characterization of Lewy body pathology in 12- and 16-year-old intrastriatal mesencephalic grafts surviving in a patient with Parkinson's disease. Mov Disord 25:1091–1096

Lippa CF, Fujiwara H, Mann DM, Giasson B, Baba M, Schmidt ML, Nee LE, O'Connell B, Pollen DA, St George-Hyslop P, Ghetti B, Nochlin D, Bird TD, Cairns NJ, Lee VM, Iwatsubo T, Trojanowski JQ (1998) Lewy bodies contain altered alpha-synuclein in brains of many familial Alzheimer's disease patients with mutations in presenilin and amyloid precursor protein genes. Am J Pathol 153:1365–1370

Liu L, Drouet V, Wu JW, Witter MP, Small SA, Clelland C, Duff K (2012) Trans-synaptic spread of tau pathology in vivo. PLoS One 7:e31302

Luk KC, Song C, O'Brien P, Stieber A, Branch JR, Brunden KR, Trojanowski JQ, Lee VM (2009) Exogenous alpha-synuclein fibrils seed the formation of Lewy body-like intracellular inclusions in cultured cells. Proc Natl Acad Sci USA 106:20051–20056

Luk KC, Kehm VM, Zhang B, O'Brien P, Trojanowski JQ, Lee VM (2012) Intracerebral inoculation of pathological alpha-synuclein initiates a rapidly progressive neurodegenerative alpha-synucleinopathy in mice. J Exp Med 209:975–986

Lundmark K, Westermark GT, Olsen A, Westermark P (2005) Protein fibrils in nature can enhance amyloid protein A amyloidosis in mice: cross-seeding as a disease mechanism. Proc Natl Acad Sci USA 102:6098–6102

McCann MJ, Rogers RC (1990) Oxytocin excites gastric-related neurones in rat dorsal vagal complex. J Physiol 428:95–108

Mendez I, Vinuela A, Astradsson A, Mukhida K, Hallett P, Robertson H, Tierney T, Holness R, Dagher A, Trojanowski JQ, Isacson O (2008) Dopamine neurons implanted into people with Parkinson's disease survive without pathology for 14 years. Nat Med 14:507–509

Meyer-Luehmann M, Coomaraswamy J, Bolmont T, Kaeser S, Schaefer C, Kilger E, Neuenschwander A, Abramowski D, Frey P, Jaton AL, Vigouret JM, Paganetti P, Walsh DM, Mathews PM, Ghiso J, Staufenbiel M, Walker LC, Jucker M (2006) Exogenous induction of cerebral beta-amyloidogenesis is governed by agent and host. Science 313:1781–1784

Morales R, Estrada LD, Diaz-Espinoza R, Morales-Scheihing D, Jara MC, Castilla J, Soto C (2010) Molecular cross talk between misfolded proteins in animal models of Alzheimer's and prion diseases. J Neurosci 30:4528–4535

Moss J, Bolam JP (2008) A dopaminergic axon lattice in the striatum and its relationship with cortical and thalamic terminals. J Neurosci 28:11221–11230

Mougenot AL, Nicot S, Bencsik A, Morignat E, Verchere J, Lakhdar L, Legastelois S, Baron T (2011) Prion-like acceleration of a synucleinopathy in a transgenic mouse model. Neurobiol Aging. doi:10.1016/j.neurobiolaging.2011.06.022

Murphy DD, Rueter SM, Trojanowski JQ, Lee VM (2000) Synucleins are developmentally expressed, and alpha-synuclein regulates the size of the presynaptic vesicular pool in primary hippocampal neurons. J Neurosci 20:3214–3220

Nemani VM, Lu W, Berge V, Nakamura K, Onoa B, Lee MK, Chaudhry FA, Nicoll RA, Edwards RH (2010) Increased expression of alpha-synuclein reduces neurotransmitter release by inhibiting synaptic vesicle reclustering after endocytosis. Neuron 65:66–79

Neumann M, Kahle PJ, Giasson BI, Ozmen L, Borroni E, Spooren W, Muller V, Odoy S, Fujiwara H, Hasegawa M, Iwatsubo T, Trojanowski JQ, Kretzschmar HA, Haass C (2002) Misfolded proteinase K-resistant hyperphosphorylated alpha-synuclein in aged transgenic mice with locomotor deterioration and in human alpha-synucleinopathies. J Clin Invest 110:1429–1439

Norris EH, Giasson BI, Lee VM (2004) Alpha-synuclein: normal function and role in neurode-generative diseases. Curr Top Dev Biol 60:17–54

Obeso JA, Rodriguez-Oroz MC, Benitez-Temino B, Blesa FJ, Guridi J, Marin C, Rodriguez M (2008) Functional organization of the basal ganglia: therapeutic implications for Parkinson's disease. Mov Disord 23(Suppl 3):S548–S559

Oueslati A, Fournier M, Lashuel HA (2010) Role of post-translational modifications in modulating the structure, function and toxicity of alpha-synuclein: implications for Parkinson's disease pathogenesis and therapies. Prog Brain Res 183:115–145

Paleologou KE, Schmid AW, Rospigliosi CC, Kim HY, Lamberto GR, Fredenburg RA, Lansbury PT Jr, Fernandez CO, Eliezer D, Zweckstetter M, Lashuel HA (2008) Phosphorylation at Ser-129 but not the phosphomimics S129E/D inhibits the fibrillation of alpha-synuclein. J Biol Chem 283:16895–16905

Papy-Garcia D, Christophe M, Huynh MB, Fernando S, Ludmilla S, Sepulveda-Diaz JE, Raisman-Vozari R (2011) Glycosaminoglycans, protein aggregation and neurodegeneration. Curr Protein Pept Sci 12:258–268

Pignatelli A, Kobayashi K, Okano H, Belluzzi O (2005) Functional properties of dopaminergic neurones in the mouse olfactory bulb. J Physiol 564:501–514

Polymeropoulos MH, Lavedan C, Leroy E, Ide SE, Dehejia A, Dutra A, Pike B, Root H, Rubenstein J, Boyer R, Stenroos ES, Chandrasekharappa S, Athanassiadou A, Papapetropoulos T, Johnson WG, Lazzarini AM, Duvoisin RC, Di Iorio G, Golbe LI, Nussbaum RL (1997) Mutation in the alpha-synuclein gene identified in families with Parkinson's disease. Science 276:2045–2047

Prusiner SB (1982) Novel proteinaceous infectious particles cause scrapie. Science 216:136–144

Prusiner SB (1998) Prions. Proc Natl Acad Sci USA 95:13363–13383

Prusiner SB (2001) Shattuck lecture – neurodegenerative diseases and prions. N Engl J Med 344:1516–1526

Simons M, Raposo G (2009) Exosomes – vesicular carriers for intercellular communication. Curr Opin Cell Biol 21:575–581

Spillantini MG, Schmidt ML, Lee VM, Trojanowski JQ, Jakes R, Goedert M (1997) Alpha-synuclein in Lewy bodies. Nature 388:839–840

Stefanis L (2012) alpha-Synuclein in Parkinson's disease. Cold Spring Harb Perspect Med 2: a009399

Steiner JA, Angot E, Brundin P (2011) A deadly spread: cellular mechanisms of alpha-synuclein transfer. Cell Death Differ 18:1425–1433

Surmeier DJ, Guzman JN, Sanchez-Padilla J, Schumacker PT (2011) The role of calcium and mitochondrial oxidant stress in the loss of substantia nigra pars compacta dopaminergic neurons in Parkinson's disease. Neuroscience 198:221–231

Vilarino-Guell C, Wider C, Ross OA, Dachsel JC, Kachergus JM, Lincoln SJ, Soto-Ortolaza AI, Cobb SA, Wilhoite GJ, Bacon JA, Behrouz B, Melrose HL, Hentati E, Puschmann A, Evans DM, Conibear E, Wasserman WW, Aasly JO, Burkhard PR, Djaldetti R, Ghika J, Hentati F, Krygowska-Wajs A, Lynch T, Melamed E, Rajput A, Rajput AH, Solida A, Wu RM, Uitti RJ, Wszolek ZK, Vingerhoets F, Farrer MJ (2011) VPS35 mutations in Parkinson disease. Am J Hum Genet 89:162–167

Volpicelli-Daley LA, Luk KC, Patel TP, Tanik SA, Riddle DM, Stieber A, Meaney DF, Trojanowski JQ, Lee VM (2011) Exogenous alpha-synuclein fibrils induce Lewy body pathology leading to synaptic dysfunction and neuron death. Neuron 72:57–71

Wakabayashi K, Takahashi H (2006) Cellular pathology in multiple system atrophy. Neuropathology 26:338–345

Wang W, Perovic I, Chittuluru J, Kaganovich A, Nguyen LT, Liao J, Auclair JR, Johnson D, Landeru A, Simorellis AK, Ju S, Cookson MR, Asturias FJ, Agar JN, Webb BN, Kang C, Ringe D, Petsko GA, Pochapsky TC, Hoang QQ (2011) A soluble alpha-synuclein construct forms a dynamic tetramer. Proc Natl Acad Sci USA 108:17797–17802

Williams JT, North RA, Shefner SA, Nishi S, Egan TM (1984) Membrane properties of rat locus coeruleus neurons. Neuroscience 13:137–156

Wood SJ, Wypych J, Steavenson S, Louis JC, Citron M, Biere AL (1999) alpha-Synuclein fibrillogenesis is nucleation-dependent. Implications for the pathogenesis of Parkinson's disease. J Biol Chem 274:19509–19512

Zimprich A, Benet-Pages A, Struhal W, Graf E, Eck SH, Offman MN, Haubenberger D, Spielberger S, Schulte EC, Lichtner P, Rossle SC, Klopp N, Wolf E, Seppi K, Pirker W, Presslauer S, Mollenhauer B, Katzenschlager R, Foki T, Hotzy C, Reinthaler E, Harutyunyan A, Kralovics R, Peters A, Zimprich F, Brucke T, Poewe W, Auff E, Trenkwalder C, Rost B, Ransmayr G, Winkelmann J, Meitinger T, Strom TM (2011) A mutation in VPS35, encoding a subunit of the retromer complex, causes late-onset Parkinson disease. Am J Hum Genet 89:168–175

Propagation and Replication of Misfolded SOD1: Implications for Amyotrophic Lateral Sclerosis

Anne Bertolotti

Abstract Amyotrophic lateral sclerosis (ALS) is a fatal and rapidly progressive motor neuron disease, with 50 % of patients dying within 1.5 years of symptoms onset. The clinical manifestations are heterogeneous in ALS, as the region of onset of muscle weakness varies between individuals. Regardless of the site of onset, the symptoms of ALS begin in one discrete body region in 98 % of the cases. Subsequently, symptoms inevitably progress to regions contiguous to the site of onset where they appear with decreasing severity. These unique clinical features suggest that neurodegeneration in ALS is an orderly and propagating process. At the molecular level, it is now well recognized that protein misfolding plays a central role in both familial and sporadic ALS. Recently, it was found that mutant SOD1, the major component of the protein deposits in familial forms of ALS, propagates misfolding from cell to cell and replicates its misfolding conformation indefinitely, just like prions do. This phenomenon could provide the molecular basis of the focality and spreading of muscle weakness in ALS, as well as the cell autonomous and non-cell autonomous processes in ALS.

Amyotrophic lateral sclerosis (ALS), first described by Jean-Martin Charcot in 1890 (Charcot 1890) is a fatal motor neuron disease with a life-time risk of 1 in 1,000. ALS is characterized by the degeneration and loss of motor neurons commencing in adult life. The resulting muscle atrophy is inevitably fatal due to respiratory failure. ALS is rapidly progressive and leads to death within 1–3 years following onset: 50 % of patients die within 1.5 years of symptoms onset, although occasionally individuals survive for more than 10 years (Kiernan et al. 2011). Likewise, prion diseases are furiously progressive, while the other neurodegenerative diseases, Alzheimer's, Parkinson's and Huntington's diseases have longer clinical course, culminating in death 10–20 years after symptoms onset.

A. Bertolotti (✉)
MRC Laboratory of Molecular Biology, Hills Road, Cambridge CB2 0QH, UK
e-mail: aberto@mrc-lmb.cam.ac.uk

M. Jucker and Y. Christen (eds.), *Proteopathic Seeds and Neurodegenerative Diseases*,
Research and Perspectives in Alzheimer's Disease,
DOI 10.1007/978-3-642-35491-5_9, © Springer-Verlag Berlin Heidelberg 2013

The clinical manifestations are heterogeneous in ALS, as the region of onset of muscle weakness varies between individuals. Bulbar onset is manifested by speech or swallowing difficulties, and limb onset with muscle weakness in the arm or leg, with the deficits usually affecting both upper and lower motor neurons (Kiernan et al. 2011). Regardless of the site of onset, the symptoms of ALS begin in one discrete body region in 98 % of the cases (Ravits et al. 2007b). Subsequently, symptoms inevitably progress to regions contiguous to the site of onset where they appear with decreasing severity. These unique clinical features initially reported by Gowers (1892) suggest that neurodegeneration in ALS is an orderly and propagating process (Ravits and La Spada 2009). Indeed, neuronal loss is usually radial in ALS with the greatest loss in the site of onset and the severity of the loss is progressively reducing in the adjacent regions (Ravits et al. 2007a).

Like Azheimer's, Parkinson's and the other neurodegenerative diseases 10 % of ALS is dominantly inherited while the majority are sporadic. In both familial and sporadic forms, proteinaceous deposits accumulate in affected neurons and glial cells and it is now well recognized that protein misfolding is a central process in both familial and sporadic forms of ALS (Lagier-Tourenne and Cleveland 2009). Mutant SOD1 aggregates in familial forms of ALS with SOD1 mutations (Valentine et al. 2005) while in most sporadic and some familial ALS, the deposits are made of TDP-43 (Neumann et al. 2006; Sreedharan et al. 2008). However, wild-type SOD1 can also be found in proteinaceous deposits in some sporadic ALS (Bosco et al. 2010).

How SOD1 deposition, a common hallmark of familial ALS (fALS), lead to motor neuron degeneration remains to be elucidated. However, it is now well established that SOD1 mutations cause fALS by a gain of toxic properties associated with the misfolding of the disease-causing protein. Mice lacking SOD1 do not develop motor neuron disease (Reaume et al. 1996). In contrast, expression of ALS-causing SOD1 mutants in transgenic mice leads to the progressive deposition of the protein and faithfully recapitulates the characteristics of the human disease: a progressive motor neuron loss leading to paralysis and death (Gurney et al. 1994; Wang et al. 2009). What elicits the progressive aggregation of mutant SOD1 in ALS is unknown.

Unlike most of the proteins associated with neurodegenerative disease, Aβ42 (Alzheimer's disease), polyQ (Huntington's disease and other polyQ disorders), Tau (Alzheimer's disease and tauopathies) and α-synuclein (Parkinson's disease), SOD1 is a highly structured protein. SOD1 is an abundant and ubiquitously expressed 32 kDa homodimeric protein which catalyses the dismutation of superoxide radicals. Each monomer folds as an eight-stranded Greek key β-barrel (Valentine et al. 2005), binds one copper and one zinc ion, and contains a disulfide bond. The native protein is amongst the most stable protein nature has designed: neither 8 M urea nor 1 % SDS dissociate the native enzyme (Malinowski and Fridovich 1979). Even more remarkably, SOD1 enzymatic activity was reported to survive >3,000 years of mummification, as an active SOD1 fragment has been recovered from an Egyptian mummy (Weser et al. 1989).

How can such an extraordinarily stable protein be converted into aggregates in ALS? More than 140, mostly missense mutations, have been reported in *SOD1* and these mutations somehow cause aggregation of the affected protein and fALS

(Valentine et al. 2005). SOD1 mutations are scattered through the entire sequence of the protein and have very diverse effects on the properties of the protein. Some mutations alter the stability of SOD1, metal binding and enzymatic activity. Intriguingly however, many mutations seem to have no measurable effect of the protein (Valentine et al. 2005). How such diverse ALS-causing SOD1 mutations all cause aggregation of the protein has been a great puzzle. Because of the diversity of the alterations provoked by SOD1 mutations, it has been assumed that different mutations cause aggregation by specific mechanisms (Shaw and Valentine 2007). However, we recently found that diverse ALS-causing mutations share a common defect: they have an increased propensity to expose normally buried hydrophobic surfaces, which drives the assembly of aggregates (Munch and Bertolotti 2010). Importantly, many SOD1 mutants which are natively folded, wild-type like, do not constitutively expose hydrophobic surfaces but have a propensity to do so when destabilised (Munch and Bertolotti 2010). These results reconciled the seemingly diverse effects of ALS-causing mutations into an unanticipated unifying mechanism. While the identification of the biochemical nature of the intermediate in the assembly of SOD1 aggregates was an important step forward, what elicits the deposition of SOD1 remained unknown.

Mutant SOD1 is expressed throughout the life of individuals with SOD1 mutations while the disease has an adult-onset with peak age around 50 years of age (Kiernan et al. 2011). Thus, the disease-causing protein must be benign for decades before acquiring some toxic properties. In transgenic mice, aggregation of an ALS-causing SOD1 mutant is an aged-dependent process (Johnston et al. 2000; Wang et al. 2009). In cells, overexpression of diverse ALS-causing SOD1 mutants does not lead to the formation of the disease characteristic deposits (Johnston et al. 2000; Munch et al. 2011). Although SOD1 mutations render the protein prone to misfold and prone to expose hydrophobic surfaces, it is soluble in cell culture, as is the case for many years in individuals with SOD1 mutations.

The key features of ALS namely the central role of protein misfolding, the variability of age of onset, the focality of muscle weakness onset with subsequent spreading of the symptoms in an orderly manner to contiguous body region have suggested the following model. The discrete site of onset may correspond to a stochastic initial misfolding of the disease-causing protein event in one cell. The spreading of the motor phenotypes may be caused by the spreading of the misfolded ALS-causing protein from the site of onset to neighbouring cells. To do so, the misfolded conformer of the disease-causing protein ought to replicate its aberrant conformation in a manner reminiscent of prions. We tested this hypothesis using SOD1.

We prepared SOD1 aggregates from highly purified recombinant proteins and labelled with fluorescent dyes (Munch and Bertolotti 2010; Munch et al. 2011). Mutant SOD1 aggregates, at a concentration of 0.2 μM monomer equivalent, were directly added to the culture media of neuronal cells. Rapidly after their inoculation, we found that aggregates penetrated in virtually all cells.

In an attempt to dissect the mechanism underlying the penetration of SOD1 aggregates into cells, we first tested whether this uptake was mediated by an active process. We found that ATP depletion dramatically reduces uptake revealing that aggregates penetrate into cells by an active mechanism. Consistently, we obtained no evidence that aggregates could directly penetrate through the plasma membrane (unpublished results). To examine whether aggregates could enter cells by an endocytic route, we investigated whether actin was required, as most endocytic pathways require changes in actin dynamics (Doherty and McMahon 2009). Uptake of SOD1 aggregates was markedly reduced by cytochalasin A or D, which prevent actin polymerisation. This suggested that SOD1 aggregates enter cells by endocytosis. Indeed, when cells were labeled with the lipophilic dye, FM 4-64FX, and inoculated with aggregates, we observed that aggregates were contained in endocytic vesicles (Munch et al. 2011).

Endocytosis occurs via different mechanisms in mammalian cells. We then systematically examined the known endocytic pathways to determine which one was required for SOD1 aggregate uptake using a combination of approaches. We inhibited selectively key components of the known endocytosis pathways using pharmacological inhibitors, dominant negative mutants or genetically modified cells when available. The thorough analysis of the different endocytic pathways revealed that extracellular mutant SOD1 aggregates uptake was not mediated by clathrin, dynamin, caveolin or flotillin. In fact, SOD1 aggregates use macropinocytosis to enter cells (Munch et al. 2011). Several viruses also highjack macropinocytosis to penetrate into the cytosol of host cell.

Following macropinocytic entry, aggregates escape micropinosomes, access the cytosol where they become ubiquitinated, just like the deposits in ALS. In the cytosol, aggregates seed aggregation of the otherwise soluble mutant SOD1 protein. Similar observations have been reported with diverse proteins associated with neurodegenerative diseases (Jucker and Walker 2011). Brain homogenates containing Aβ or tau induce aggregation of the then soluble mouse peptide or protein respectively (Kane et al. 2000; Meyer-Luehmann et al. 2006; Clavaguera et al. 2009). When directly inoculated onto cells aggregates made of polyQ, tau or synuclein seed aggregation of their intracellular counterparts (Ren et al. 2009; Frost et al. 2009; Danzer et al. 2007, 2009; Desplats et al. 2009; Hansen et al. 2011; Volpicelli-Daley et al. 2011).

The most remarkable and defining property of prions is their ability to replicate their aberrant conformation indefinitely. This property was so far only reported for prions. Strikingly, we found that after it has been induced, mutant SOD1 aggregation propagates indefinitely (Fig. 1). It is not yet known whether the other human proteins associated with neurodegenerative diseases replicate their misfolded conformers indefinitely (Munch and Bertolotti 2012). Future studies will most certainly clarify this critical point.

Following transient exposure of cells with exogenous SOD1 aggregates, prepared from a highly purified protein, we have induced a new phenotype in cells as they chronically display a misfolding pathology that is maintained long after the disappearance of the seeds (Munch et al. 2011). This is reminiscent of a protein-only mode of inheritance where information is transmitted through a self-replicating conformation of a protein (Munch and Bertolotti 2011). Furthermore, we found that mutant SOD1 aggregates transfer from cell to cell with high efficiency, a process that does not require contacts between cells but depends on the extracellular release of aggregates. These findings reveal that SOD1 shares the *self-replicating* property and transmissibility of prions.

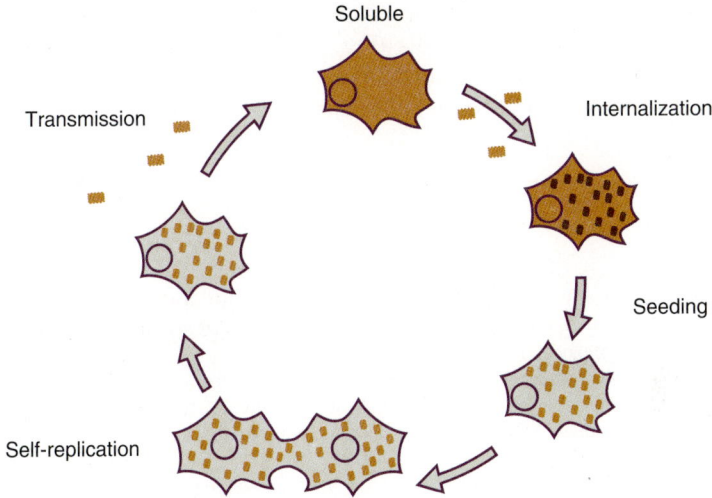

Fig. 1 Self-replication of mutant SOD1 misfolding: the vicious cycle. SOD1 aggregates enter cells by macropinocytosis and rapidly access the cytosol the convert the otherwise cellular protein into aggregates. This is a point of no return. Once it has been induced, aggregation of SOD1 replicates itself indefinitely and is transmitted to neighbouring cells upon the extracellular release of aggregates

In addition to our observations suggesting a role for extracellular SOD1 aggregates in ALS, extracellular SOD1 has been detected in cerebrospinal fluid of ALS patients (Jacobsson et al. 2001). It has also been reported that SOD1 can be secreted, although it is primarily cytosolic (Mondola et al. 2003; Urushitani et al. 2006). The recent success of SOD1 immunotherapy applied to transgenic mice are compatible with a pathophysiological role of extracellular SOD1 in ALS (Urushitani et al. 2007).

It has long been known that the damages in ALS are determined by both glia and neurons in ALS mouse models (Boillee et al. 2006; Clement et al. 2003). Remarkably, in chimeric mice containing are a mixture of normal and SOD1 mutant cells, wild-type neurons acquire abnormal ubiquitination when adjacent to SOD1 mutant expressing cells (Clement et al. 2003). These non-cell autonomous features of ALS have been recapitulated in vitro with the transmission of toxicity form ALS-astrocytes to wild-type motor neurons in culture (Nagai et al. 2007; Haidet-Phillips et al. 2011). Surprisingly, knockdown of SOD1 expression in sALS astrocytes reduced their toxicity to motor neurons. It remains to be investigated whether the agent that transfers toxicity between neuronal and non-neuronal cells is misfolded SOD1 itself.

Conclusion

The observations described here indicate that at the cellular and molecular level, misfolded SOD1 mutant share the defining properties of prions: the ability to replicate their abnormal conformer indefinitely. The vicious-cycle by which

SOD1 propagates misfolding from cell to cell and its ability to replicate its misfolding conformation indefinitely (Fig. 1) could provide the molecular basis of the focality and spreading of muscle weakness in ALS, the cell autonomous and non-cell autonomous processes in ALS and the success of immunotherapy in mice.

References

Boillee S, Yamanaka K, Lobsiger CS, Copeland NG, Jenkins NA, Kassiotis G, Kollias G, Cleveland DW (2006) Onset and progression in inherited ALS determined by motor neurons and microglia. Science 312:1389–1392

Bosco DA, Morfini G, Karabacak NM, Song Y, Gros-Louis F, Pasinelli P, Goolsby H, Fontaine BA, Lemay N, McKenna-Yasek D, Frosch MP, Agar JN, Julien JP, Brady ST, Brown RHJ (2010) Wild-type and mutant SOD1 share an aberrant conformation and a common pathogenic pathway in ALS. Nat Neurosci 13:1396–1403

Charcot JM (1890) Lecons sur les Maladies due Systeme Nerveux, 3 vols. Bureau du Progres Medical, Paris, 1894

Clavaguera F, Bolmont T, Crowther RA, Abramowski D, Frank S, Probst A, Fraser G, Stalder AK, Beibel M, Staufenbiel M, Jucker M, Goedert M, Tolnay M (2009) Transmission and spreading of tauopathy in transgenic mouse brain. Nat Cell Biol 11:909–913

Clement AM, Nguyen MD, Roberts EA, Garcia ML, Boillee S, Rule M, McMahon AP, Doucette W, Siwek D, Ferrante RJ, Brown RHJ, Julien JP, Goldstein LS, Cleveland DW (2003) Wild-type nonneuronal cells extend survival of SOD1 mutant motor neurons in ALS mice. Science 302:113–117

Danzer KM, Haasen D, Karow AR, Moussaud S, Habeck M, Giese A, Kretzschmar H, Hengerer B, Kostka M (2007) Different species of alpha-synuclein oligomers induce calcium influx and seeding. J Neurosci 27:9220–9232

Danzer KM, Krebs SK, Wolff M, Birk G, Hengerer B (2009) Seeding induced by alpha-synuclein oligomers provides evidence for spreading of alpha-synuclein pathology. J Neurochem 111: 192–203

Desplats P, Lee HJ, Bae EJ, Patrick C, Rockenstein E, Crews L, Spencer B, Masliah E, Lee SJ (2009) Inclusion formation and neuronal cell death through neuron-to-neuron transmission of alpha-synuclein. Proc Natl Acad Sci USA 106:13010–13015

Doherty GJ, McMahon HT (2009) Mechanisms of endocytosis. Annu Rev Biochem 78:857–902

Frost B, Jacks RL, Diamond MI (2009) Propagation of tau misfolding from the outside to the inside of a cell. J Biol Chem 284:12845–12852

Gowers WR (1892) Manual of diseases of the nervous system, vol I. J. & A. Churchill, London

Gurney ME, Pu H, Chiu AY, Dal Canto MC, Polchow CY, Alexander DD, Caliendo J, Hentati A, Kwon YW, Deng HX et al (1994) Motor neuron degeneration in mice that express a human Cu, Zn superoxide dismutase mutation. Science 264:1772–1775

Haidet-Phillips AM, Hester ME, Miranda CJ, Meyer K, Braun L, Frakes A, Song S, Likhite S, Murtha MJ, Foust KD, Rao M, Eagle A, Kammesheidt A, Christensen A, Mendell JR, Burghes AH, Kaspar BK (2011) Astrocytes from familial and sporadic ALS patients are toxic to motor neurons. Nat Biotechnol 29:824–828

Hansen C, Angot E, Bergstrom AL, Steiner JA, Pieri L, Paul G, Outeiro TF, Melki R, Kallunki P, Fog K, Li JY, Brundin P (2011) alpha-Synuclein propagates from mouse brain to grafted dopaminergic neurons and seeds aggregation in cultured human cells. J Clin Invest 121: 715–725

Jacobsson J, Jonsson PA, Andersen PM, Forsgren L, Marklund SL (2001) Superoxide dismutase in CSF from amyotrophic lateral sclerosis patients with and without CuZn-superoxide dismutase mutations. Brain 124:1461–1466

Johnston JA, Dalton MJ, Gurney ME, Kopito RR (2000) Formation of high molecular weight complexes of mutant Cu, Zn-superoxide dismutase in a mouse model for familial amyotrophic lateral sclerosis. Proc Natl Acad Sci USA 97:12571–12576

Jucker M, Walker LC (2011) Pathogenic protein seeding in Alzheimer disease and other neurodegenerative disorders. Ann Neurol 70:532–540

Kane MD, Lipinski WJ, Callahan MJ, Bian F, Durham RA, Schwarz RD, Roher AE, Walker LC (2000) Evidence for seeding of beta-amyloid by intracerebral infusion of Alzheimer brain extracts in beta-amyloid precursor protein-transgenic mice. J Neurosci 20:3606–3611

Kiernan MC, Vucic S, Cheah BC, Turner MR, Eisen A, Hardiman O, Burrell JR, Zoing MC (2011) Amyotrophic lateral sclerosis. Lancet 377:942–955

Lagier-Tourenne C, Cleveland DW (2009) Rethinking ALS: the FUS about TDP-43. Cell 136:1001–1004

Malinowski DP, Fridovich I (1979) Subunit association and side-chain reactivities of bovine erythrocyte superoxide dismutase in denaturing solvents. Biochemistry 18:5055–5060

Meyer-Luehmann M, Coomaraswamy J, Bolmont T, Kaeser S, Schaefer C, Kilger E, Neuenschwander A, Abramowski D, Frey P, Jaton AL, Vigouret JM, Paganetti P, Walsh DM, Mathews PM, Ghiso J, Staufenbiel M, Walker LC, Jucker M (2006) Exogenous induction of cerebral beta-amyloidogenesis is governed by agent and host. Science 313:1781–1784

Mondola P, Ruggiero G, Seru R, Damiano S, Grimaldi S, Garbi C, Monda M, Greco D, Santillo M (2003) The Cu, Zn superoxide dismutase in neuroblastoma SK-N-BE cells is exported by a microvesicles dependent pathway. Mol Brain Res 110:45–51

Munch C, Bertolotti A (2010) Exposure of hydrophobic surfaces initiates aggregation of diverse ALS-causing superoxide dismutase-1 mutants. J Mol Biol 399:512–525

Munch C, Bertolotti A (2011) Self-propagation and transmission of misfolded mutant SOD1: prion or prion-like phenomenon? Cell Cycle 10:1711

Munch C, Bertolotti A (2012) Propagation of the prion phenomenon: beyond the seeding principle. J Mol Biol 421(4–5):491–498

Munch C, O'Brien J, Bertolotti A (2011) Prion-like propagation of mutant superoxide dismutase-1 misfolding in neuronal cells. Proc Natl Acad Sci USA 108:3548–3553

Nagai M, Re DB, Nagata T, Chalazonitis A, Jessell TM, Wichterle H, Przedborski S (2007) Astrocytes expressing ALS-linked mutated SOD1 release factors selectively toxic to motor neurons. Nat Neurosci 10:615–622

Neumann M, Sampathu DM, Kwong LK, Truax AC, Micsenyi MC, Chou TT, Bruce J, Schuck T, Grossman M, Clark CM, McCluskey LF, Miller BL, Masliah E, Mackenzie IR, Feldman H, Feiden W, Kretzschmar HA, Trojanowski JQ, Lee VM (2006) Ubiquitinated TDP-43 in frontotemporal lobar degeneration and amyotrophic lateral sclerosis. Science 314:130–133

Ravits JM, La Spada AR (2009) ALS motor phenotype heterogeneity, focality, and spread: deconstructing motor neuron degeneration. Neurology 73:805–811

Ravits J, Laurie P, Fan Y, Moore DH (2007a) Implications of ALS focality: rostral-caudal distribution of lower motor neuron loss postmortem. Neurology 68:1576–1582

Ravits J, Paul P, Jorg C (2007b) Focality of upper and lower motor neuron degeneration at the clinical onset of ALS. Neurology 68:1571–1575

Reaume AG, Elliott JL, Hoffman EK, Kowall NW, Ferrante RJ, Siwek DF, Wilcox HM, Flood DG, Beal MF, Brown RHJ, Scott RW, Snider WD (1996) Motor neurons in Cu/Zn superoxide dismutase-deficient mice develop normally but exhibit enhanced cell death after axonal injury. Nat Genet 13:43–47

Ren PH, Lauckner JE, Kachirskaia I, Heuser JE, Melki R, Kopito RR (2009) Cytoplasmic penetration and persistent infection of mammalian cells by polyglutamine aggregates. Nat Cell Biol 11:219–225

Shaw BF, Valentine JS (2007) How do ALS-associated mutations in superoxide dismutase 1 promote aggregation of the protein? Trends Biochem Sci 32:78–85

Sreedharan J, Blair IP, Tripathi VB, Hu X, Vance C, Rogelj B, Ackerley S, Durnall JC, Williams KL, Buratti E, Baralle F, de Belleroche J, Mitchell JD, Leigh PN, Al-Chalabi A, Miller CC, Nicholson G, Shaw CE (2008) TDP-43 mutations in familial and sporadic amyotrophic lateral sclerosis. Science 319:1668–1672

Urushitani M, Sik A, Sakurai T, Nukina N, Takahashi R, Julien JP (2006) Chromogranin-mediated secretion of mutant superoxide dismutase proteins linked to amyotrophic lateral sclerosis. Nat Neurosci 9:108–118

Urushitani M, Ezzi SA, Julien JP (2007) Therapeutic effects of immunization with mutant superoxide dismutase in mice models of amyotrophic lateral sclerosis. Proc Natl Acad Sci USA 104:2495–2500

Valentine JS, Doucette PA, Zittin Potter S (2005) Copper-zinc superoxide dismutase and amyotrophic lateral sclerosis. Annu Rev Biochem 74:563–593

Volpicelli-Daley LA, Patel TP, Luk KC, Tanik SA, Riddle DM, Stieber A, Meaney DF, Trojanowski JQ, Lee VM (2011) Exogenous alpha-synuclein fibrils induce Lewy body pathology leading to synaptic dysfunction and neuron death. Neuron 72:57–71

Wang J, Farr GW, Zeiss CJ, Rodriguez-Gil DJ, Wilson JH, Furtak K, Rutkowski DT, Kaufman RJ, Ruse CI, Yates JR, Perrin S, Feany MB, Horwich AL (2009) Progressive aggregation despite chaperone associations of a mutant SOD1-YFP in transgenic mice that develop ALS. Proc Natl Acad Sci USA 106:1392–1397

Weser U, Miesel R, Hartmann HJ (1989) Mummified enzymes. Nature 341:696

Development of Drugs That Target Proteopathic Seeds Will Require Measurement of Drug Mechanism in Human Brain

Peter T. Lansbury Jr

Abstract The notion that many neurodegenerative diseases are caused by seeded protein aggregation is almost 20 years old (Jarrett and Lansbury, Cell 73:1055–1058, 1993; Lansbury, Neuron 19:1151–1154, 1997). Recent data, some of it summarized here, suggest that this mechanism may account for cell-to-cell transmission throughout the brain by "proteopathic seeds." There are many scientific questions that remain to be solved, including what is the best approach to interfere with the seeding process in vivo. But it may be more important to address the practical bottleneck that is common to all therapeutic strategies: how can one demonstrate potential efficacy of an experimental drug in a small, inexpensive clinical trial? This manuscript will address the issues that, together, have produced this bottleneck and will suggest some possible approaches to stimulate drug development for neurodegeneration.

An important question for everyone with a stake in public health to ask is, "given the explosion of scientific research into the underlying cause of Alzheimer's disease (AD) over the past 20–25 years, why isn't there a disease-modifying drug, that is, a drug that slows disease progression, on the horizon?" Such a drug would change the course of many lives, possibly changing AD to a manageable health problem that would not necessarily progress to complete debilitation and institutionalization (Fig. 1b as compared to Fig. 1a). This change would save tens of billions of public health insurance dollars per year. A disease-modifying drug for AD would likely become one of the best-selling drugs of all time, probably topping $15B in annual sales. So why is the pharmaceutical industry drastically deemphasizing their own efforts to find such a drug (Petsko 2011)?

The explanation for the current drought is primarily an economic one, based on the aversion in the twenty-first century pharmaceutical industry to take on any

P.T. Lansbury Jr (✉)
Center for Neurologic Diseases, Brigham and Women's Hospital, 77 Avenue Louis Pasteur, Boston, MA 02115, USA
e-mail: peter@linkmedicine.com

M. Jucker and Y. Christen (eds.), *Proteopathic Seeds and Neurodegenerative Diseases*, Research and Perspectives in Alzheimer's Disease, DOI 10.1007/978-3-642-35491-5_10, © Springer-Verlag Berlin Heidelberg 2013

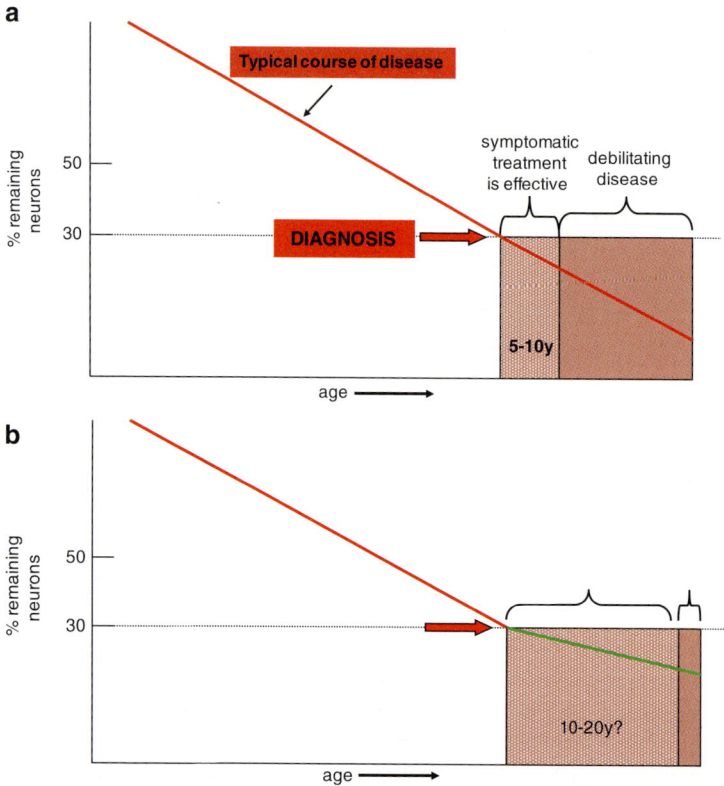

Fig. 1 (**a**) Progression of Alzheimer's disease. Alzheimer's disease typically progresses very slowly and does not produce symptoms until many neurons have been lost. (**b**) A disease-modifying drug would reduce the prevalence of debilitating disease

short-term risk at the expense of long-term innovation. There is an empirical basis for this situation. A March 2012 study from the Tufts University Center for the Study of Drug Development (DiMasi and Grabowski 2012) analyzed the record of the pharmaceutical industry between 1996 and 2010 and determined that CNS drugs required 35 % longer from initiation of clinical trials to approval than drugs for non-CNS indications. Moreover, these drugs were much less likely to succeed (10 % vs. 17 % success rate). The impact of these facts has been amplified in the wake of the failures of recent clinical trials. However, certain intrinsic characteristics of AD present real scientific challenges to drug development. These challenges can be solved by scientific innovation, but the risk-averse pharmaceutical industry has relegated this effort to the academic community. In order for the fantastic science, some of which is reported here, to ever impact human beings, it is critical that the community producing this science understands and addresses the issues listed below.

Tailoring drugs to act in the brain provides unique challenges that increase the cost and time of development

The brain is a protected organ, so any drug needs to penetrate the blood-brain barrier and avoid rapid efflux by the PgP system, which has as its function the removal of foreign toxins from the brain. This issue puts tremendous pressure on medicinal chemists; it is solvable, but introduces limitations as to what molecular scaffolds can be utilized. The combined difficulties in targeting proteins in the brain have resulted in most pharmaceutical companies insisting on demonstration of "target engagement" by drug in the human brain. This step typically requires synthesis of a PET ligand and adds to the time and cost of development. However, if possible, this is a desirable goal in that it allows for selection of dose at an early stage.

The progression of AD symptoms is extremely slow and irregular, making it extremely expensive to obtain evidence supporting the efficacy of a particular strategy

Thus, a clinical trial in which it is critical to detect a clear difference between the placebo group and the treatment group, must run for 18 months (Fig. 2) and involve many patients. This type of trial typically takes 3 years from start to finish and costs between $100M and $300M, an enormous sum to invest in an untested strategy (of course, all novel strategies are untested at this point). This first trial in patients takes on such importance because there are no validated animal models of AD (of course, this is a circular argument; the first clinical success will break this logjam), so the human trial is the first real indication of whether a particular strategy will show efficacy. Most mouse models are designed with a particular therapeutic strategy in mind (e.g., reduction of amyloid plaque) and are designed to demonstrate a clear phenotype within a year. Thus, models based on amyloid precursor protein (APP) overexpression have been produced in which the level is much larger than is seen in all but a few families with early-onset AD. These mice produce amyloid plaques but do not suffer from neuronal loss. In some cases, but only at certain ages (Reed et al. 2010), cognitive deficits are measurable. Therapeutic strategies based on amyloid reduction have delivered impressive results in these mice, reducing plaque and improving cognition. These mice have proved to be reliable predictors of amyloid reduction in the clinic, but cognitive improvement in mice has never translated into cognitive improvement in humans. In fact, in the case of the Eli Lilly gamma secretase inhibitor, reduction of plaque and improvement of cognition in a mouse model translated into decreased cognition in a human trial (Sperling et al. 2011; Selkoe 2011).

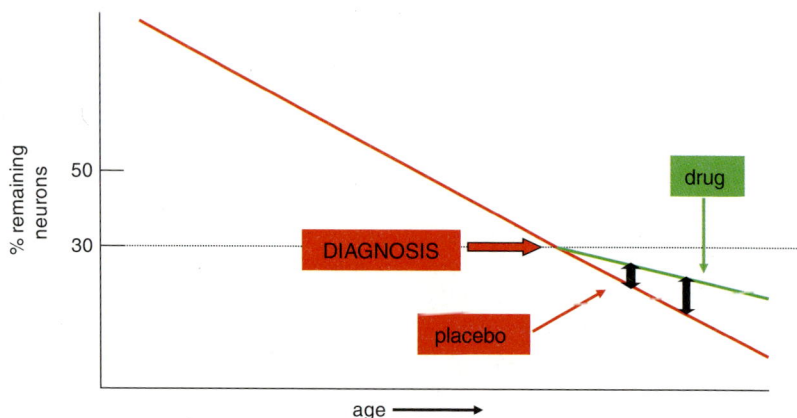

Fig. 2 Demonstration of drug efficacy will require a very long and/or large (i.e., expensive) clinical trial. The gap between the placebo (*red*) and drug-treated (*green*) groups will grow as the trial runs. To detect this gap, a large number of patients must be included

Due to the difficulty in testing hypotheses regarding disease etiology in predictive animal models, it is unclear whether pathological correlates of disease progression are cause, effect, or epiphenomenon

Therapeutic strategies based on pathological features of disease (e.g., amyloid plaque) are inherently risky and cannot easily be tested. These pathological markers could be driving disease, of course, but could also represent an endogenous response to disease or even an epiphenomenon unrelated to the disease process. Take for example, amyloid plaques, which, thanks to Alzheimer's original description of the disease, are perfectly correlated with AD (that is, every AD patient has brain amyloid plaques). However, it is now clear that many aged individuals who are cognitively normal have extensive amyloid plaques. Furthermore, in cognitively normal individuals aged 90–100, amyloid load and cognitive ability are not correlated (Savva et al. 2009; Maarouf et al. 2011). Finally, amyloid plaque load, as measured by PET imaging, does not correlate with symptom severity (Boche et al. 2010). Nonetheless, reduction of amyloid plaque has been utilized as a surrogate for efficacy by several companies trying to develop amyloid vaccination or passive immunization approaches. In all cases, reduction of amyloid plaque, as measured by PET imaging, was achieved by several of these approaches and this effect was used to justify moving into large phase 3 trials. However, no change in cognition was detected in those same patients. The failure of plaque and cognition to correlate in small phase 2 trials calls into question the original assumption that plaques are a cause of AD. Despite these issues, the targeting of amyloid plaque is now in the process of being clinically tested. These studies should, once and for all, test whether amyloid plaque (1) causes AD or (2) is an adaptive response to the disease or an epiphenomenon of underlying causative defects. The time and cost of getting to the point where a single therapeutic strategy is testable have been enormous. Whether or not these initial studies are successful, it is unlikely that comparable resources will be invested in the testing of any other hypothesis in the near future.

One must develop methods to monitor the underlying processes that are being targeted by a new experimental drug

These methods will enable the execution of a confidence-building (or risk-reducing) bridge to a more expensive disease-modifying trial. The difficulty is which process to target: first, neither the pathological features (plaques, tangles) of AD nor the protein products of causative genes (presenilin [PS], APP) have been clearly linked to a particular cellular process, so it is unclear how to monitor the effects of compounds directed at these targets. Second, AD is a disease of the brain, so the response to an experimental drug cannot be monitored by biopsying the diseased tissue, as it is in oncology, for example. Brain imaging is extremely sophisticated and can be used to monitor drug response, but most current imaging efforts are directed at the development of imaging markers that correlate to symptom progression. These markers are critical, but they measure effects that are downstream of the ideal response marker. I propose that increased effort be directed at elucidating the processes that underlie disease and developing imaging probes that directly measure these processes. Such probes can be used for early stage clinical trials, where it will be critical to demonstrate that the drug modulates a relevant process in the brain in a short period of time. This type of data will be required to justify the risk and expense of a trial aimed at demonstrating slowed disease progression.

The single most important issue: it will be critical to carefully select patients for early clinical trials, since it is unlikely that response to any drug will be uniform

Since it will be no longer be possible to jump from safety trials right into an 18-month disease modifying trial, it will be absolutely critical to demonstrate signs of efficacy (e.g., activation of a disease-associated defective process in brain) in cost-effective early trials. The selection of patients for these early trials is essential to their success; it will be important to optimize chances for success using a subpopulation before moving to larger "all-comers" trials. The criteria for selection will include stage of disease; early patients are considered likely to be more responsive than late patients. There has been a lot of discussion of the need to diagnose AD at an earlier stage than current guidelines allow. This discussion has as its premise that it may only be possible to affect disease modification if treatment begins presymptomatically (Sperling et al. 2011). Although it is clearly true that the disease process is well underway at the point when symptoms meet the criteria for diagnosis, there is no evidence that presymptomatic treatment will have a better chance of working [in mouse models of AD, treatment regimens that modify pathogenesis, e.g., plaque formation, do not necessarily work best when administered early in the process (Townsend et al. 2010)]. Furthermore, it is unlikely that any pharmaceutical company would test an experimental drug with any significant side effects in pre-symptomatic patients unless some efficacy had first been demonstrated in symptomatic patients. Exceptions may be made in the case of approved drugs; for example, the diabetes drug pioglitazone is being tested by Takeda in a 5-year trial in patients with prodromal AD (Takeda press release 2011). Thus it is not realistic to expect a wave of presymptomatic drug trials unless and until one can obtain some evidence of efficacy in human patients. Therefore, other approaches to increase the odds of observing efficacy must be explored.

One such approach involves developing methods to select patients that are likely to respond to a given drug. Since most cases of AD are not likely to be caused by a single underlying defect, the notion of a "silver bullet" drug that is widely effective is unrealistic. Most AD patients likely suffer from a combination of defects that contribute to their disease. The degree to which each defect contributes will determine the degree to which each individual responds to a drug targeting that particular defect. Thus it will be critical to develop methods to characterize the "flavor" of AD of every individual so as to be able to select a subpopulation to populate an optimal trial. It is an interesting question as to whether the heterogeneity of AD, which has been widely ignored in the pharmaceutical industry, has caused the effectiveness of certain experimental drugs to go undetected (an exception is apoE genotype, which is being used to stratify patients for anti-amyloid trials, although the mechanistic rationale for doing so is unclear). There are examples of this situation in oncology; take, for example, Pfizer's recently approved drug for a subpopulation of non small-cell lung cancer patients. This drug was originally tested in an unselected population and was deemed a failure. When a mutation was fortuitously discovered in the gene encoding the drug's target kinase, a second trial, involving only those patients bearing the mutation, was designed and efficacy became clear. AD is likely more complex than most oncology indications in that many defects contribute. Therefore, phenotypic sub-typing of AD patients may make more sense that a genotypic subtyping. This problem will receive more attention in the coming years.

Can we target "proteopathic seeds"?

It is useful to consider how the above issues impact the translation of the scientific findings summarized in this symposium to the clinic. First off, while there is good evidence that disease pathology can be spread from cell to cell in the brain, it is not yet clear that this spread is necessarily correlated to spread of neuronal cell loss

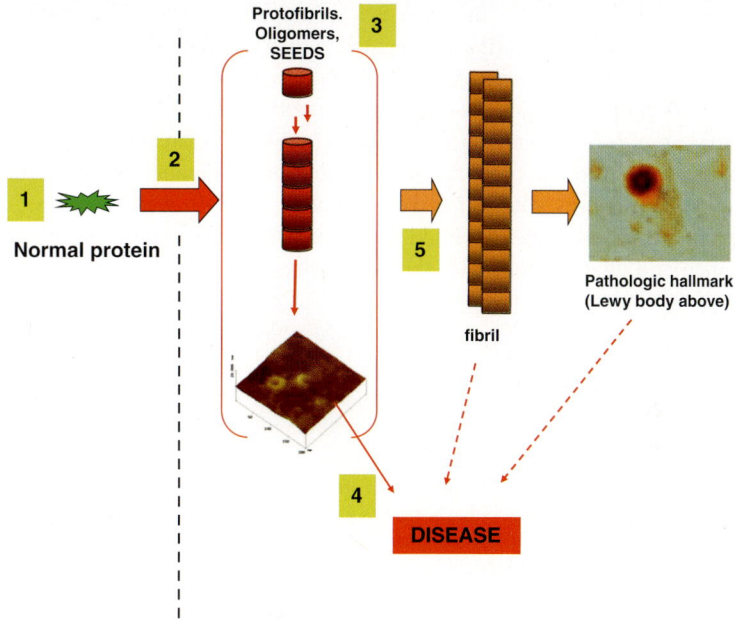

Fig. 3 When normal protein crosses a threshold for aggregation (*dotted line*), disease ensues. Formation of fibrillar aggregates is likely to represent a "detoxification" mechanism. Possible therapeutic strategies include (*1*) reducing protein expression, (*2*) inhibiting protein aggregation, (*3*) promoting aggregate degradation, (*4*) inhibiting aggregate toxicity, and (*5*) promoting formation of inert fibrillar aggregates

(this correlation has been convincingly demonstrated in mouse models of prion disease). Furthermore, the mechanism linking protein aggregation to cell death remains to be elucidated. It may be that several mechanisms are responsible. Thus, it makes much more sense to target events upstream of aggregate toxicity (Fig. 3); (1) production of aggregating protein, (2) formation of protein seed, (3) degradation of protein seed, and (4) intercellular "transmission" of protein seed. Of these, the first is much less attractive, since the aggregating proteins presumably have biological functions that could be compromised by decreased expression. Furthermore, an expression-based strategy would have to be specific for a single protein, since a general inhibitor of transcription/translation would have toxicity problems. Likewise, prevention of intercellular "transmission" seems to be too risky at this time, since the mechanism is not understood and it is not clear that transmission is required for disease propagation. The two strategies that seem to be viable and generalizable across various protein aggregation indications involve the prevention of seed formation (by a small molecule or by an induced chaperone) and promotion of seed degradation (by promotion of autophagy, which targets protein aggregates and not functional monomers). Both of these strategies are being widely pursued.

So, while the scientific prospects are encouraging, solutions to the issues of translation raised above will determine whether or not drugs are developed. Can methods be developed to selectively measure the proteopathic seed in the human brain (as opposed to the inert fibrillar aggregates, which are measured by existing PET imaging agents)? Is it possible to measure the activity of the autophagy machinery, which targets protein aggregates for degradation, in brain? Which patients are most likely to respond to the anti-seed approach (familial Parkinson's disease patients characterized by protein overexpression, such as the synuclein triplication and duplication, or patients with defective autophagy; how do we find them?). All of these problems are solvable, provided that the scientific community recognizes their importance and attacks them with the same intensity as has been applied towards studies of the underlying biology.

References

Boche D, Denham N, Holmes C, Nicoll JA (2010) Neuropathology after active Aβ42 immuno-therapy: implications for Alzheimer's disease pathogenesis. Acta Neuropathol 120:369–384

DiMasi JA, Grabowski HG (2012) R&D costs and returns to new drug development: a review of the evidence. In: Danzon PM, Nicholson S (eds) Tufts Center for the Study of Drug Development. March/April CSDD Impact Report. The Oxford handbook of the economics of the biopharmaceutical industry. Oxford University Press, Oxford, pp 21–46

Jarrett JT, Lansbury PT (1993) Seeding the "one-dimensional crystallization" of amyloid: a pathogenic mechanism in Alzheimer's disease and scrapie? Cell 73:1055–1058

Lansbury PT (1997) Structural neurology: are seeds at the root of neuronal degeneration? Neuron 19:1151–1154

Maarouf CL, Daugs ID, Kokjohn TA, Walker DG, Hunter JM, Kruchowsky JC, Woltjer R, Kaye J, Castaño EM, Sabbagh MN, Beach TG, Roher AE (2011) Alzheimer's disease and non-demented high pathology control nonagenerians: comparing and contrasting the biochemistry of cognitively successful aging. PLoS One 6:e27291

Petsko GA (2011) Bailing out. Genome Biol 12:131–133

Reed MN, Liu P, Kotilinek LA, Ashe KH (2010) Effect size of reference memory deficits in the Morris water maze in Tg2576 mice. Behav Brain Res 212:115–120

Savva GM, Wharton SB, Ince PG, Forster G, Matthews FE, Brayne C (2009) Medical Research Council Cognitive Function and Ageing Study. Age, neuropathology, and dementia. N Engl J Med 360:2302–2309

Selkoe DJ (2011) Resolving controversies on the path to Alzheimer's therapeutics. Nat Med 17:1060–1065

Sperling RA, Jack CR, Aisen PS (2011) Testing the right target and right drug at the right stage. Sci Transl Med 3:111

Takeda press release (2011) http://www.takeda.com/press/article_39958.html

Townsend M, Qu Y, Gray A, Wu Z, Seto T, Hutton M, Shearman MS, Middleton RE (2010) Oral treatment with a gamma-secretase inhibitor improves long-term potentiation in a mouse model of Alzheimer's disease. J Pharmacol Exp Ther 333:110–119

The Role of Functional Prions in the Persistence of Memory Storage

Eric R. Kandel, Irina Derkatch, and Elias Pavlopoulos

Abstract Cellular and molecular studies of both implicit and explicit memory suggest that experience-dependent modulation of synaptic strength and structure is a fundamental mechanism by which these memories are encoded and stored within the brain. Implicit and explicit memory share in common several molecular steps and an overall molecular logic. Both have two general stages are created in at least two stages: a short-term phase that does not require the synthesis of new protein and a long-term phase that does. Short-term memory involves covalent modification of preexisting proteins and changes in the strength of preexisting synaptic connections, while long-term memory requires transcriptional activation mediated by CREB and the growth of new connections. The distinction between short- and long-term memory and the recruitment of a CREB-mediated activation of gene expression leading to the growth of new synaptic connections have turned out to be almost universal. Maintenance of long-term memory involves, in addition, the functional prion CPEB that regulates local protein synthesis. The self-sustaining, prion-like, activation of CPEB3 appears to be a quite general mechanism for the perpetuation of memory.

Memory is not a unitary faculty of mind but exists in at least two forms: implicit and explicit. Implicit memory refers to memory for perceptual and motor skills. Explicit memory is the conscious recall of knowledge about facts and events—for people, places and objects. Studies of both implicit and explicit memory have found that each of these forms has two general stages: a short-term phase lasting minutes and a long-term memory phase lasting days, weeks, or longer. Studies of implicit memory storage in *Aplysia* and explicit memory storage in mice have revealed that these two distinct stages in behavioral memory are reflected in distinct stages of synaptic

E.R. Kandel (✉) • I. Derkatch • E. Pavlopoulos
Center for Neurobiology and Behavior, Howard Hughes Medical Institute, Columbia University, 1051 Riverside Dr, Room 668 Annex, New York, NY 10032, USA
e-mail: erk5@columbia.edu

M. Jucker and Y. Christen (eds.), *Proteopathic Seeds and Neurodegenerative Diseases*, 131
Research and Perspectives in Alzheimer's Disease,
DOI 10.1007/978-3-642-35491-5_11, © Springer-Verlag Berlin Heidelberg 2013

plasticity and underlying molecular mechanisms. Short-term memory involves covalent modification of pre-existing proteins and a functional change of pre-existing connections, whereas long-term memory involves CREB-mediated activation of transcription and the growth of new synaptic connections. Nearly every form of long-term memory that has been examined recruits CREB-mediated transcription as at least one component of the molecular switch from short- to long-term memory.

Retrograde Signaling from the Activated Synapses to the Nucleus

The requirement for long-term memory of CREB-mediated transcription in the nucleus, an organelle that is in contact with all the synapses of a neuron, might cause one to expect that the long-term synaptic change would have to be cell-wide. However, experiments using local applications of serotonin to a small subset of synapses in the *Aplysia* bifurcated sensory neuron-two motor neuron culture preparation (Martin et al. 1997; Casadio et al. 1999), as well as parallel experiments by Frey and Morris in the hippocampus (1997), demonstrated that specific synapses of a neuron could be modified independently in a protein synthesis-dependent manner. Thus, long-term facilitation (LTF) and the associated synaptic changes are synapse-specific, implying that there is a retrograde signaling from the activated synapses to the nucleus and subsequent anterograde signaling from the nucleus to the activated synapses.

How do the activated synapses signal to the nucleus for the initiation of memory-related gene transcription? Thompson et al. (2004) have recently found that serotonin stimulation of a synapse, which produces LTF in *Aplysia* sensory-motor neuron co-cultures, triggers the nuclear translocation of importins, proteins involved in carrying cargoes through nuclear pore complexes. Similarly, NMDA activation or long-term potentiation (LTP) induction, but not depolarization, leads to translocation of importin in hippocampal neurons (Thompson et al. 2004). Although details underlying the translocation of these retrograde signals remain unknown, the effector molecules identified thus far appear to be conserved in both invertebrates and vertebrates. Future identification of the molecular cargoes of importin and its signaling role in the nucleus are likely to increase our understanding of how transcription-dependent memory is regulated.

Anterograde Signaling from the Nucleus to the Activated Synapses: The "Synaptic Capture" Hypothesis

Following transcriptional activation, newly synthesized gene products, both mRNAs and proteins, have to be delivered to the synapses whose activation originally triggered the wave of gene expression. To explain how this specificity can be achieved in a biologically economical way in spite of the massive number of synapses in a single neuron, Martin et al. (1997) and Frey and Morris (1997) proposed the "synaptic capture" hypothesis. This hypothesis, also referred as "synaptic tagging," proposes that the products of gene expression are delivered throughout the cell to all of its synapses but are only functionally incorporated in those synapses that have been tagged by previous synaptic activity. The "synaptic tag" model has been supported by a number of studies both in the rodent hippocampus (Frey and Morris 1997, 1998; Barco et al. 2002; Dudek and Fields 2002) and *Aplysia* (Martin et al. 1997; Casadio et al. 1999).

Molecular Mechanisms of Synaptic Capture

Studies of synaptic capture at the synapses between the sensory and motor neurons of the gill-withdrawal reflex in *Aplysia* have demonstrated that the achievement of synapse-specific LTF requires more than the activation of CREB-driven gene transcription in the nucleus. Injection of phosphorylated CREB-1 into the cell body of the sensory neuron gives rise to LTF at all of the synapses by providing them with the gene products of CRE-driven genes. This facilitation, however, is not maintained beyond 24–48 h; neither is it accompanied by synaptic growth, which is achieved only if the synapse is also marked by the short-term process, a single pulse of serotonin (Casadio et al. 1999).

How is a synapse marked? Kelsey Martin found two distinct components of marking in *Aplysia* (Fig. 1). The first component initiates long-term synaptic plasticity and growth and requires the cAMP-dependent protein kinase A (PKA). The second component stabilizes long-term functional and structural changes at the synapse and requires (in addition to protein synthesis in the cell body) local protein synthesis at the synapse (Martin et al. 1997).

Since mRNAs are made in the cell body, the need for the local translation of some mRNAs suggests that these mRNAs are likely to be dormant before they reach the activated synapse. If that were true, one way of activating protein synthesis at the activated synapse would be to recruit a regulator of translation at this synapse that is capable of activating dormant mRNAs.

Therefore, Kausik Si began to search for such a regulator of protein synthesis. Experiments in *Xenopus* oocytes, by Joel Richter, had found that—in these oocytes—maternal RNA was silent until activated by a regulator of protein synthesis, the cytoplasmic polyadenylation element binding protein (CPEB; Richter 1999). Si searched for a homolog in *Aplysia* and found—in addition to the developmental isoform described by Richter—a new isoform of CPEB (ApCPEB) with novel properties. ApCPEB had four important features that made it an attractive candidate for a synapse-specific mark for stabilization: (1) it was activated through an extracellular signal; (2) it was spatially restricted (Bally-Cuif et al. 1998; Schroeder et al. 1999; Tan et al. 2001); (3) it activated mRNAs that were translationally dormant (Stebbins-Boaz et al. 1996) and (4) some of the mRNA targets of CPEB were involved in cellular growth (Chang et al. 2001; Groisman et al. 2002). Si found that

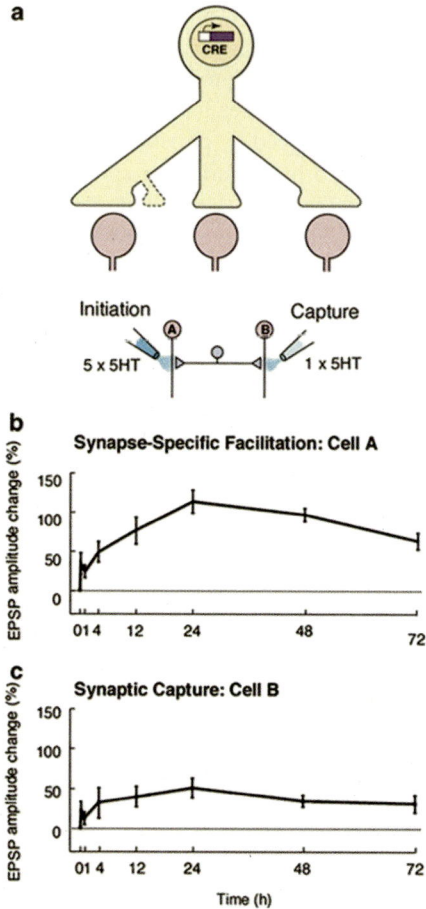

Fig. 1 Synapse-specific facilitation and synaptic capture. (**a**) Synapse-specific memory storage is a cell biological paradox. All the synapses of a neuron share a single nucleus. One would expect that the recruitment of the nucleus for the long-term process would result in cell-wide facilitation. Contrary to this notion, memory storage is synapse-specific. In this illustration, a single neuron makes connections with three different target neurons. However, only connections with one neuron are strengthened due to stimulation. How then is synapse-specific memory storage achieved? Is it possible to have delivery of molecules only to the stimulated synapses or is there a mechanism that exists at the synapse whereby only stimulated synapses can utilize the delivered molecules? (**b**) In the bifurcated sensory neuron, application of five pulses of 5HT leads to synapse-specific facilitation lasting several days. This facilitation is sensitive to disruption by transcriptional and translational inhibitors. (**c**) Synapse-specific facilitation can be captured and maintained by the other branch with the application of a single pulse of 5HT (Martin et al. 1997; Casadio et al. 1999; Puthanveettil and Kandel 2011)

blocking this isoform at a marked (active) synapse prevented the maintenance but not the initiation of synaptic LTF (Si et al. 2003a, b). Indeed, blocking the local synthesis of ApCPEB selectively blocks learning-related synaptic growth and the persistence of LTF a few days after they are formed (Miniaci et al. 2008). In addition, rapamycin, which inhibits translation of a specific set of mRNAs, selectively blocks the maintenance phase of long-term plasticity when locally applied to a synapse (Casadio et al. 1999). Consistent with this finding and also with the idea that ApCPEB might be part of the synaptic mark for the stabilization of LTF and growth, ApCPEB increases in response to stimulation by serotonin and this increase is blocked by rapamycin (Si et al. 2003a).

A Prion-Like Mechanism of CPEB Regulates Local Protein Synthesis and Thereby Maintains Synaptic Growth Through Cytoplasmic Polyadenylation

The maintenance of long-term memories poses a problem because there is a constant turnover of proteins at the synapse. The general solution to this problem of molecular turnover presumably lies in a class of stable and self-sustaining biochemical reactions.

Since proteins have a relatively short half-life compared to the duration of memory, structural changes at the synaptic level were postulated to confer stability to the memory, and it was implicitly assumed that the requirement of activity-dependent molecular changes was transient. However, it was soon clear that the maintenance of learning-related structural alterations required ongoing macromolecular synthesis (Kandel 2001). Hence, in 1984, Crick first addressed the possibility of a sustained molecular alteration as the basis of long-term memory storage, using protein phosphorylation as a candidate mechanism. In response to this suggestion, John Lisman (Lisman et al. 1997) developed a model based on the autocatalytic properties of CamKII. According to Lisman's model, synaptic stimulation activates CamKII, which can then convert inactive CamKII molecules to their active form in the absence of any further synaptic input.

Based on the properties of *Aplysia* neuronal CPEB, as a rapamycin-sensitive mark, Si et al. (2003a, b, 2010) began to explore its properties in great detail. Whereas the developmental CPEB from *Xenopus* is activated through phosphorylation (Mendez et al. 2000), no consensus phosphorylation sites were revealed in the neuronal isoform of ApCPEB.

One clue for stable and self-perpetuating chemical reactions came from the analysis of amino acid sequences of ApCPEBs. Kausik Si observed that the neuronal isoform of ApCPEB had a glutamine-asparagine (Q and N)-rich N-terminal domain similar to yeast prions. A search of the protein sequence database revealed putative homologs of the *Aplysia* neuronal CPEB in *Drosophila*, mouse, and human, with N-terminal extensions of similar character. The presence of a Q/N-rich N-terminal domain of neuronal CPEB typical of yeast prion proteins led Si et al. (2003b) to explore whether ApCPEB had prion-like properties (Fig. 2b).

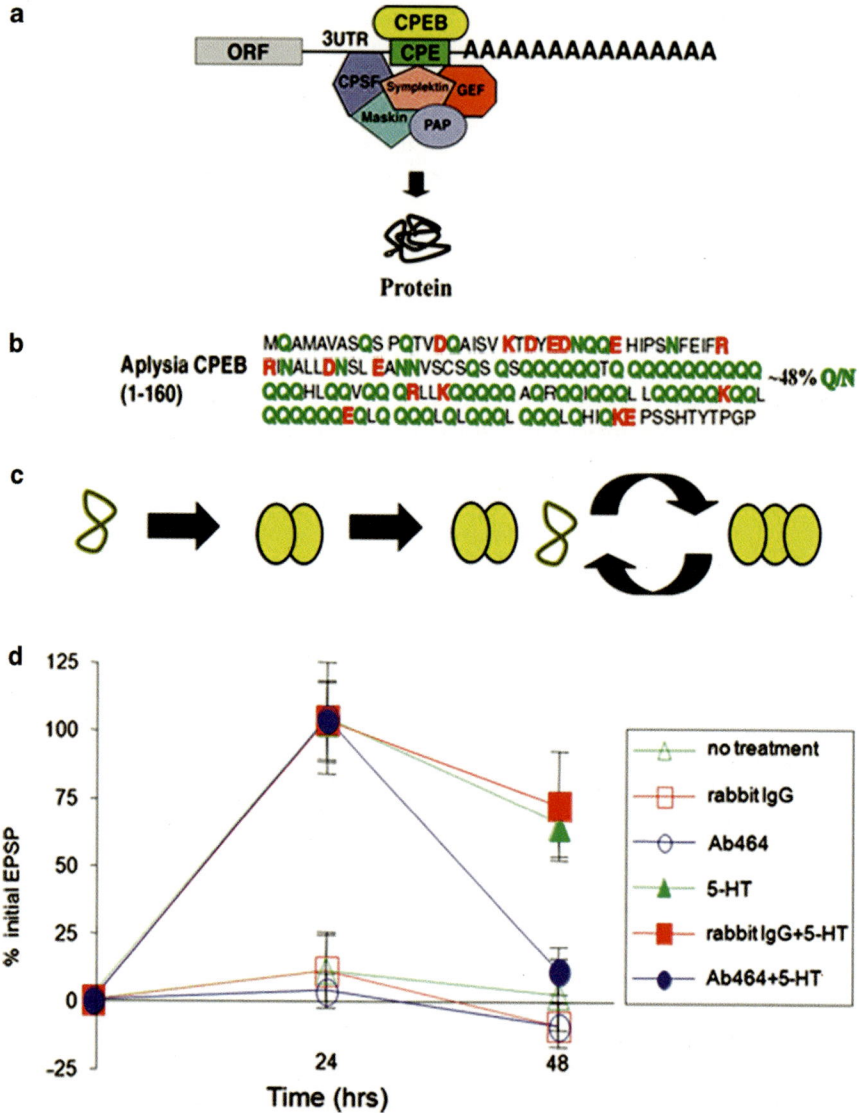

Fig. 2 The neuronal isoform of *Aplysia* has prion-like properties. (**a**) CPEB is a translational regulator and binds to mRNAs containing CPE and regulates cytoplasmic polyadenylation. This polyadenylation activates translationally repressed RNAs. (**b**) *Aplysia* CPEB has a Q/N-rich N-terminal domain that is similar to the prion domain of yeast prion. (**c**) Self-perpetuation of ApCPEB through a prion-like mechanism. Application of 5HT to sensory motor neuron synapses converts CPEB in inactive conformation to active conformation. This activated CPEB further converts inactive forms to active forms, leading to self-perpetuation of local polyadenylation. (**d**) Inhibition of formation of active aggregates of ApCPEB in sensory neurons by microinjection of specific antibody (Ab464) blocks persistence of LTF (Si et al. 2010; Puthanveettil and Kandel 2011)

The prion hypothesis suggests that biological information can be maintained indefinitely through self-perpetuating conformations of proteins. Although it was discovered in the context of Creutzfeldt-Jakob disease—a neurodegenerative disease in mammals (Griffith 1967; Prusiner 1982)—the prion-like protein mechanism has since grown to encompass non-Mendelian traits in fungi (Wickner 1994; Crow and Li 2011; Shorter and Lindquist 2005). Various prions are relatively benign and, in some cases, can confer selectable advantages to the organism (True et al. 2004; True and Lindquist 2000; Coustou et al. 1997; Eaglestone et al. 1999). The self-templating property of prions makes them epigenetically dominant and enables prion-forming proteins to function as metastable cellular switches.

To explore the possible prion-like properties of the *Aplysia* neuronal CPEB, Si et al. (2003b) used an array of assays developed for studying prions in yeast and found that ApCPEB could exist in yeast in two distinct conformational states, very much like other prion proteins (Fig. 2c). One of these states is monomeric, inactive and not capable of self-perpetuation. The other state is multimeric, active, self-perpetuating, heritable and stable across generations in yeast, although occasional switches occur (Si et al. 2003b). Furthermore, the fusion of the N-terminal domain of ApCPEB to the developmental isoform, which is normally not self-perpetuating, was capable of triggering a self-perpetuating change in state with the properties of a yeast prion.

Si et al. (2010) then went on beyond yeast to explore the conformational state of ApCPEB in neurons. They found that, in *Aplysia* sensory neurons when ApCPEB was overexpressed, it formed punctate structures that were amyloid in nature, a common characteristic of all known prions. Using a fluorescence reconstitution assay in which the two halves of GFP were attached to ApCPEB monomers, Si et al. found that these punctate structures are formed by multimerization of ApCPEB monomers. Intriguingly, the application of five pulses of serotonin, which produces LTF, increased the number of puncta, suggesting that the multimerization of ApCPEB3 was regulated by modulating synaptic activity. Moreover, injection of an antibody that selectively binds the multimeric form of ApCPEB did not prevent the initiation of LTF but selectively blocked its maintenance beyond 24 h (Fig. 2d). Photoconvertible GFP experiments further demonstrated that serotonin treatment was capable of recruiting new ApCPEB proteins to pre-existing multimers.

An activity-dependent prion-like switch could therefore serve as a mechanism to maintain a self-sustained activated molecular state. According to this model, ApCPEB in the sensory neuron has at least two conformational states: (1) a recessive monomeric state where ApCPEB3 is inactive or acts as a repressor of translation, and (2) a dominant, self-sustaining, active multimeric state. In a naive synapse, the basal level of ApCPEB is low and the protein is in the monomeric state. An increase in the amount of ApCPEB induced by 5-HT, either by itself or in conjunction with other signals, results in the conversion of ApCPEB from the monomeric to the prion-like state, which might be more active or be devoid of the inhibitory function of the basal state. Once the prion state is established at an activated synapse, dormant mRNAs, made in the cell body and distributed globally to all synapses, can be activated only locally by means of cytoplasmic polyadenylation through the

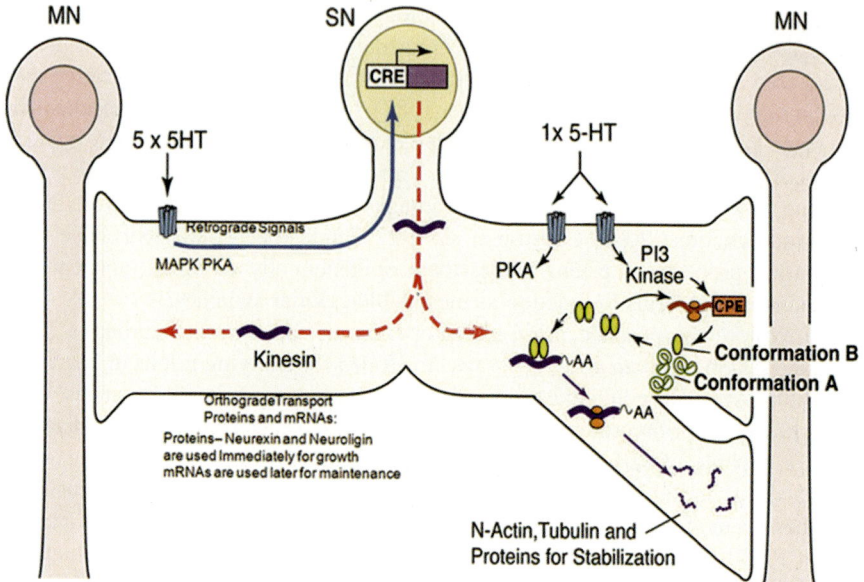

Fig. 3 A model for the initiation and persistence of long-term memory storage. A single pulse of 5HT to sensory neuron (SN) and motor neuron (MN) synapses recruits the short-term process, and this requires PKA activity. PI3 kinase also becomes activated and stimulates CPEB-dependent translation. Five pulses of 5HT activate CREB in the nucleus through the PKA-MAPK pathway and facilitate enhanced, kinesin-mediated fast axonal transport of proteins, mRNAs and organelles to synapses. These activation steps are critical for the initiation of LTF. RNAs transported by kinesin may be used for persistence. During the persistence phase, CPEB at the stimulated synapse activates polyadenylation of CPE-containing RNAs through a prion-like mechanism for the self-perpetuation of synapse-specific memory storage (Puthanveettil and Kandel 2011)

activated ApCPEB. Because the activated ApCPEB can be self-perpetuating, it could contribute to a self-sustaining, synapse-specific, long-term molecular change and provide a mechanism for the stabilization of learning-related synaptic growth and the persistence of memory storage (Fig. 3).

An interesting feature of ApCPEB is that, unlike most other prions that once converted into their multimeric state, are pathogenic or, at best, dead proteins, ApCPEB appears to be a functional prion. The active, self-perpetuating form of the protein does not kill cells but rather has an important physiological function.

However, the involvement of a prion-like molecule in the formation of synapse-specific, long-term memory raises several issues. How is the autocatalytic, self-perpetuating chemical reaction resulting in synaptic enhancement regulated? One possibility is that molecular chaperones might regulate the conformational states of ApCPEB. Indeed Si et al. (2003b) found that overexpression of the heat shock protein HSP104 is capable of reversing the active form of ApCPEB in yeast. A second consideration is that the action of ApCPEB should be highly restricted to specific synapses: if ApCPEB in the active state could escape from a newly potentiated synapse, the entire cellular ApCPEB would then be converted to an active state, thereby leading to erroneous potentiation of all synapses. Thus, there must be a mechanism that effectively restricts the activated ApCPEB at the potentiated synapse. In support of such regulation, Si et al. (2010) found that the activated ApCPEB auto-associates and polymerizes, thus restricting its diffusion.

Compartmentalization of activated ApCPEB could also be facilitated through interactions of its polymers with other cellular components, including those that are translated upon ApCPEB activation. For example, actin cytoskeleton is involved in the formation, propagation and intracellular localization of the yeast prion $[PSI^+]$ (Ganusova et al. 2006), and oligomeric ApCPEB has been shown to activate translation of actin mRNAs (Si et al. 2003a).

The critical role and mode of action of CPEB in the maintenance of implicit long-term memory are not restricted to *Aplysia*. The Si lab and the Barry Dickson lab have found, independently, that long-term memory in *Drosophila* also involves CPEB (Mastushita-Sakai et al. 2010; Keleman et al. 2007). Si and Dickson similarly found that a learned courtship behavior, in which males are conditioned to suppress their courtship upon prior exposure to unreceptive females, is not suppressed when the prion domain of the *Drosophila* CPEB, orb2a, is deleted. As a result, there is loss of long-term courtship memory. The Si lab has recently demonstrated that the amyloid-like oligomers of orb2a are critical for the persistence of long-term memory in the fly (Majumdar et al. 2012).

Explicit Memory Storage and the Cytoplasmic Polyadenylation Element Binding Protein-3

Is the function of CPEB in regulating and maintaining long-lasting changes in synaptic plasticity evolutionary conserved? Is CPEB also critical for the maintenance of explicit long-term memory storage in mammals? To answer these questions, we searched for CPEB orthologs in the mouse and found four isoforms; we have called these isoforms CPEB-1 to CPEB-4 (Theis et al. 2003). One of these isoforms, CPEB-3, is most similar to ApCPEB (Fig. 4). Joseph Stephan et al. found that CPEB-3 has prion-like properties in yeast (Stephan et al. unpublished observations), and Elias Pavlopoulos found that it is highly regulated in the mouse brain.

Pavlopoulos found that CPEB-3 was activated by Neuralized1, an E3 ubiquitin ligase (Pavlopoulos et al. 2011). He found that CPEB-3 colocalized with Neuralized1 in dendrites of adult hippocampal neurons and in opposition to the presynaptic marker Synaptophysin (Fig. 5). Moreover, Neuralized1 and CPEB-3 physically interacted in the adult hippocampus (Fig. 5). In mice overexpressing Neuralized1 in the forebrain, the levels of CPEB-3 were increased in the hippocampus, whereas CPEB-1 and CPEB-4 were unaffected (Fig. 6). By using a reduced system of cultured hippocampal neurons, Pavlopoulos found that CPEB-3 interacted with Neuralized1 via its N-terminal, prion-like domain, and that this interaction led to the monoubiquitination and consequent activation of CPEB-3 (Fig. 7). He observed that, in the basal state,

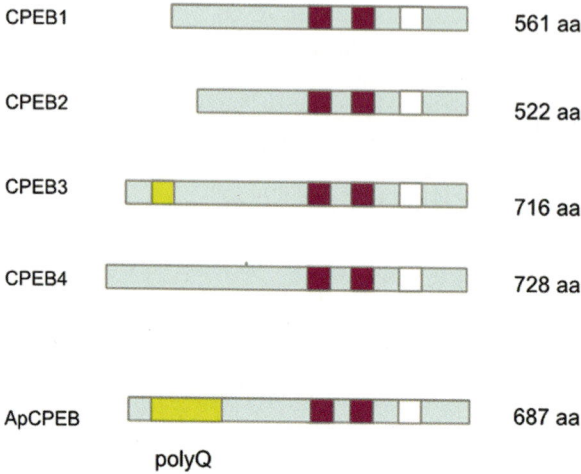

Fig. 4 Mice have four isoforms of CPEB: CPEB1- to CPEB-4. The N-terminal domain of CPEB-3 resembles the prion-like domain of ApCPEB

CPEB-3 was a negative regulator of the translation of its target mRNAs, GluA1 and GluA2 subunits of AMPA type receptors, and that this action of CPEB-3 was reversed by activity that induced upregulation of Neuralized1 and direct ubiquitination of CPEB-3. Strikingly, overexpression of Neuralized1 mimics synaptic activity and activates CPEB-3 in cultured hippocampal neurons.

Consistent with the data in cultured hippocampal neurons, conditional overexpression of Neuralized1 in the adult mouse forebrain also increases the relative amount of monoubiquitinated CPEB-3 and the polyadenylation, translation, and protein levels of GluA1 and GluA2. Mice overexpressing Neuralized1 have increased numbers of postsynaptic dendritic spines and functional synapses and enhanced hippocampus-dependent memory and synaptic plasticity. By contrast, inhibition of Neuralized1 reduces the steady state levels of monoubiquitinated CPEB-3, leading to the shortening of the poly(A) tails, reduced translation of GluA1 and GluA2 mRNAs, and interference with hippocampus-dependent memory and synaptic plasticity (Pavlopoulos et al. 2011).

These results suggest a model whereby Neuralized1-mediated ubiquination facilitates hippocampal plasticity and hippocampus-dependent memory storage by modulating the activity of CPEB-3 and CPEB-3-dependent protein synthesis and synapse formation. According to this model, Neuralized1 turns over rapidly in the basal state because it is degraded by the proteasome (Fig. 8). During this state, CPEB-3 acts as a repressor of translation. In response to synaptic activity (Fig. 8), the protein levels of Neuralized1 are increased, and Neuralized1 translocates and moves in proximity to CPEB-3, leading to the ubiquination and activation of CPEB-3 and consequent production of synaptic components critical for the formation of new, functional synaptic connections.

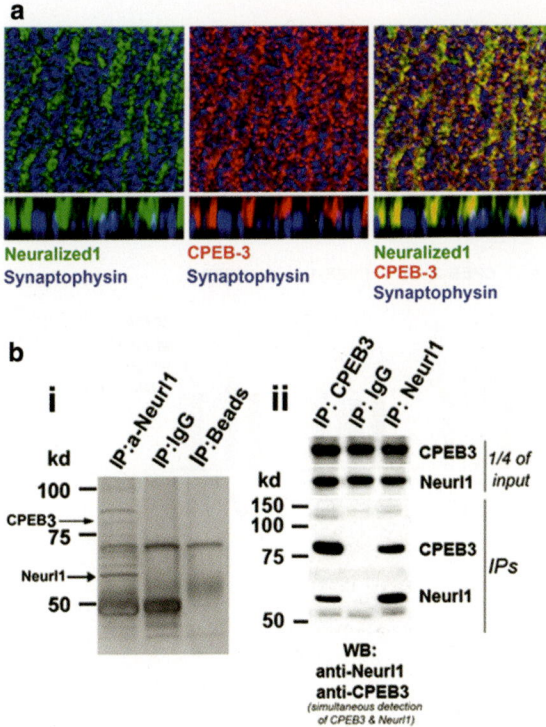

Fig. 5 The E3 ubiquitin ligase Neuralized1 interacts with CPEB-3 in the hippocampus. (**a**) Confocal images of the dendritic layer of CA1 pyramidal neurons. Neuralized1 is co-localized with CPEB-3 dendrites and both are in apposition to a presynaptic (synaptophysin) site in the hippocampus. *Upper panel*: magnification of apical dendrites. *Lower panel*: cross sections. (**b**) (**i**) Silver staining of hippocampal lysates from adult wild type mice after immunoprecipitation with anti-Neuralized1 antibody. CPEB3 was detected as Neuralized1-interacting protein by mass spectrometry. (**ii**) Co-immunoprecipitation experiments from hippocampal lysates of adult wild type mice. CPEB-3 and Neuralized1 participate in the same macromolecular complex (Pavlopoulos et al. 2011)

To test this hypothesis directly, Pavlopoulos et al. (2011) carried out experiments in cultured hippocampal neurons. They found that, in the control state, there was a certain number of spines (Fig. 9a). If CPEB-3 was then overexpressed in these cells, it led to a repression of translation and significantly reduced the number of spines (Fig. 9b). However, when Neuralized1 was overexpressed in the neurons so that it could monoubiquinate the overexpressed CPEB-3, the number of spines returned to the baseline state (Fig. 9c). If the prion domain of CPEB-3 was eliminated, spine number was reduced (Fig. 9d). When CPEB-3 was overexpressed together with Neuralized1 lacking its ligase activity, the number of dendritic spines also was reduced (Fig. 9e). These results demonstrate that Neuralized1-mediated monoubiquitination of CPEB-3 modulates CPEB-3 activity and suggest that the monoubiquitinated form of CPEB-3 is the active form of the protein. Pavlopoulos

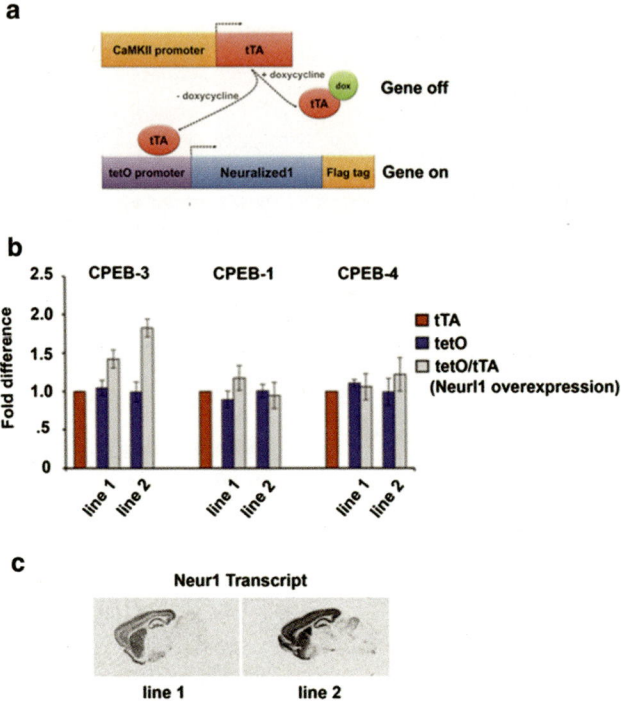

Fig. 6 In two mouse lines overexpressing Neuralized1 (Neurl) in the adult forebrain, the levels of CPEB-3 are increased systematically but the levels of CPEB-1 and CPEB-4 are not. (**a**) Schematic representation of the tetO/tTA system. In the double transgenic mice (tetO/tTA), tTA binds to the tetO promoter and activates the expression of the Neuralized1 (Neurl1) transgene only in the forebrain where the CaMKIIα promoter is active. The expression of the transgenes is also temporarily regulated in a reversible manner, because the binding of the tTA transactivator to the tetO promoter is inhibited by tetracycline or its analogs (doxycycline; dox). (**b**) Averaged data from 36 mice. Mean fold difference + SEM is shown. CPEB3 is specifically increased in Neurl1 overexpressing mice in a dosage-dependent manner. CPEB-1 and CPEB-4 levels are unaffected. (**c**) In situ hybridization analysis of Neurl1 mRNA on sagittal brain sections from adult double transgenic mice (Pavlopoulos et al. 2011)

et al. addressed this issue directly by expressing in cultured hippocampal neurons CPEB-3 that was artificially monoubiquitinated. They found that monoubiquitinated CPEB-3 without Neuralized1 could, by itself, lead to the growth of new synaptic spines (Fig. 9f).

The mechanism of CPEB-3 regulation that Pavlopoulos et al. demonstrated is consistent with the idea that CPEB-3 is the mammalian homolog of *Aplysia* CPEB and that CPEB-3 regulates local protein synthesis and is required for the stabilization of synaptic growth and long-lasting change of synaptic efficacy—much as we found for ApCPEB. The proposed mechanism is also consistent with the idea that CPEB-3, similar to ApCPEB, is an explosive molecule; therefore, its function has to be tightly regulated so that it can respond rapidly to synaptic stimulation and control mRNA translation at the proper time and at specific, activated, synaptic sites.

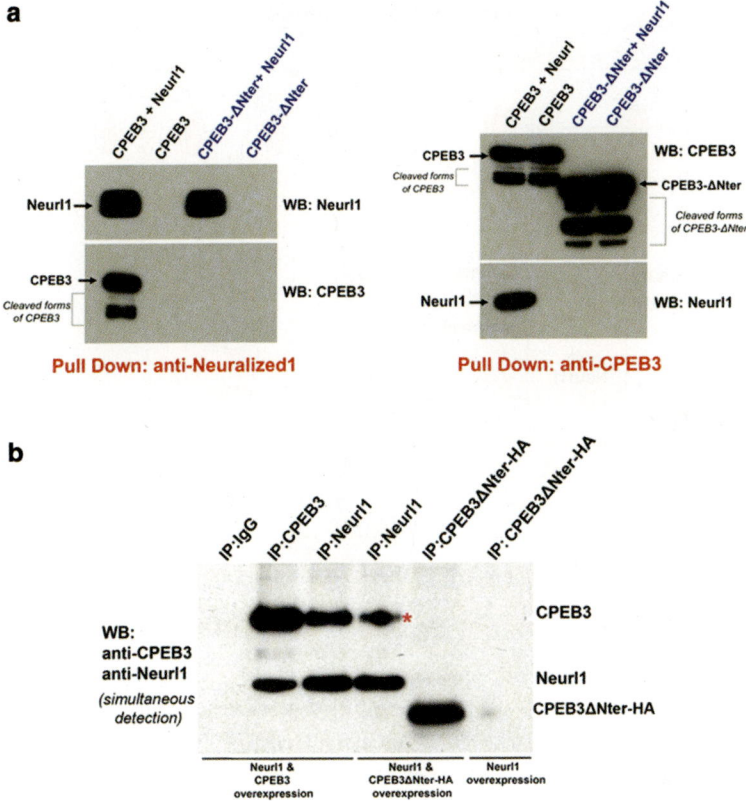

Fig. 7 The N-terminal (prion-like) domain of CPEB-3 is critical for the interaction of CPEB-3 with Neuralized1. Co-immunoprecipitation experiments in either mammalian cell line (**a**) (HEK293 cells) or in cultured hippocampal neurons (**b**). (**b**) CPEB3 lacking its N-terminal domain was fused to the HA epitope tag (CPEB3ΔNter-HA) so that it could be discriminated from the endogenous CPEB3. Anti-HA antibody was used for IP of CPEB3-ΔNter-HA. *Red asterisk*: Neurl1 co-immunoprecipitates endogenous CPEB3 but not CPEB3ΔNter-HA (Pavlopoulos et al. 2011)

CPEB Represents a New Class of Functional Prions

There are several features of the prion-like state of ApCPEB and CPEB-3 that are similar to and several that are novel and distinctly different from both the pathological PrP^{Sc} protein and the nonpathological but inactive state of yeast prions. First, although ApCPEB and CPEB-3 form a large number of multimers in *Aplysia* and hippocampal neurons, these multimers, like yeast prions, have no adverse effect either on basal synaptic functions or on survival of the neurons over days. Second, the results with the CPEB homologs suggest that, unlike other known prions that are inactive, these CPEBs retain biochemical activity in their multimeric state, such as the ability to bind mRNA. However, it is not yet evident whether these are active sites of poly(A) tail elongation or protein synthesis. Finally, in all known prions, the conversion to the

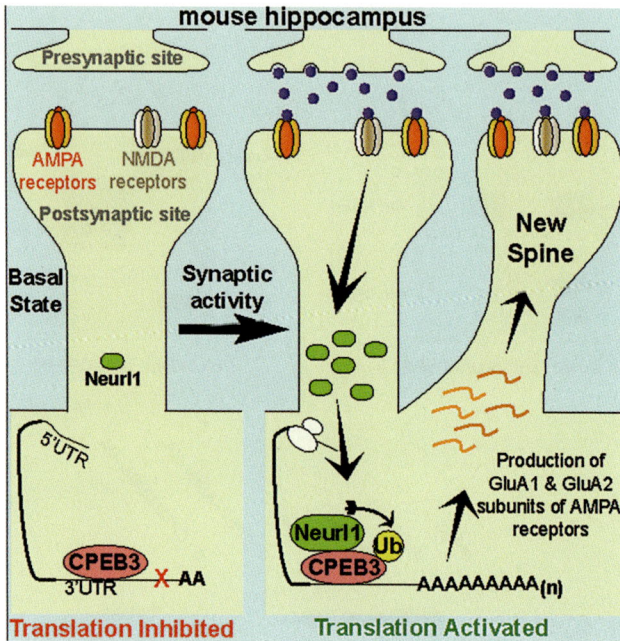

Fig. 8 The E3 ubiquitin ligase Neuralized1 activates CPEB-3. Neuralized1-mediated ubiquination facilitates hippocampal plasticity and hippocampus-dependent memory storage by modulating the activity of CPEB-3 and CPEB-3-dependent protein synthesis and synapse formation

prion state is spontaneous. In contrast, conversion of ApCPEB and CPEB-3 to the multimeric state is regulated by a physiological signal.

In addition to self-maintenance, multimerization of the prion-like CPEBs might lead to a net increase in activity by increasing the number of active units. If the monomeric form of the proteins has weak activity, having multiple units would effectively increase activity. There might be other, more interesting possibilities. The prion-like multimerization requires a significant conformational conversion that, in addition to preserving the original activity, might confer new activities to the protein.

Self-perpetuating conformational reactions, reminiscent of prion conversion to the "infectious scrapie" form, might underlie the capacity of ApCPEB to self-perpetuate memory in an epigenetic fashion. The initial activation of ApCPEB begins a local positive feedback loop within the synapse induced by learning, whereby ApCPEB promotes the activation of more ApCPEB through specific conformational changes, which is potentially a very powerful mechanism for generating a self-perpetuating local signal at the synapse.

Recently, another example was added to the collection of functional prions. Hou et al. (2011) showed that MAVS, a mitochondrial protein that activates the transcription factors IRF3 and NF-kappa B to induce type 1 interferons, forms functional prion-like aggregates to activate and propagate anti-viral innate immune

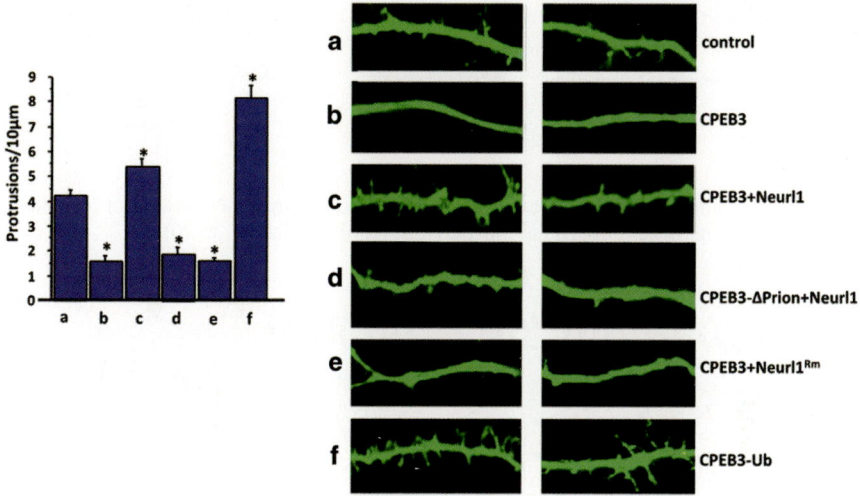

Fig. 9 The modulation of CPEB3 by Neuralized1 (Neurl1) and ubiquitin alters the number of spines in cultured hippocampal neurons. Modulation is blocked by removal of the prion or ubiquitin ligase domains. Dendrites of cultured hippocampal neurons expressing EGFP alone (control) or EGFP and the indicated proteins are shown. The averaged density of spines $\pm$ SEM is also shown (Pavlopoulos et al. 2011)

responses. They found that viral infection induces the formation of very large MAVS aggregates, which potently activate IRF3. The MAVS fibrils behave like prions and effectively convert endogenous MAVS into functional aggregates. MAVS is activated by the RIG-I-like RNA helicases. RIG-I binds to unanchored lysine-63 (K63) polyubiquitin chains, in addition to its binding to the viral RNA, and this binding is important for MAVS activation. Interestingly, Hou et al. found that the presence of unanchored K63 polyubiquitin chains was required for RIG-I to catalyze the conversion of MAVS on the mitochondrial membrane to prion-like aggregates. These results, in combination with our data, trigger the intriguing question about whether ubiquitin and ubiquitin-like molecules may play a central role in regulating the aggregation state of functional prions, their activity, or both.

Functional Prions Are Likely to Have a Distinctive Structure That Can Be Regulated

Conventional prions have a beta-sheet-rich structure, and the structural transition from soluble to aggregated forms occurs through uncontrolled misfolding and is unlikely to undergo regulated structural transitions (Fig. 10). With functional prions, such as *Aplysia* CPEB and CPEB-3, the conversion from one state to another is regulated by physiological signals. Thus, Ferdinando Fiumara searched for other

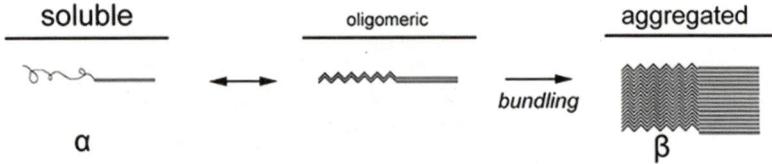

Fig. 10 Conventional prions have a beta sheet structure and the structural transitions from soluble to aggregated forms occur through uncontrolled misfolding (Fiumara et al. 2010)

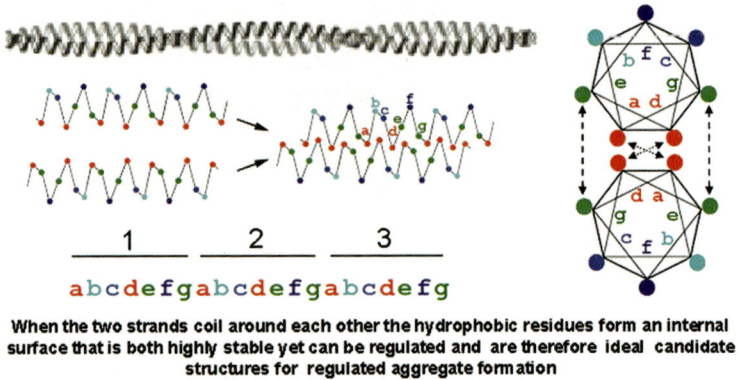

abcdefgabcdefgabcdefg

When the two strands coil around each other the hydrophobic residues form an internal surface that is both highly stable yet can be regulated and are therefore ideal candidate structures for regulated aggregate formation

Fig. 11 Coiled-coil alpha helices can mediate oligomerization, like beta sheets. Moreover, unlike beta sheets, coiled-coils can change their oligomeric state in response to signals from the environment. Coiled-coils have a periodic pattern of heptad repeats with hydrophobic residues in the a and d positions often occupied by leucine, isoleucine, methionine or valine. In addition, glutamine can substitute for hydrophobic residues in the a and d position in N/Q rich prions and in ApCPEB. When the two strands coil around each other, positions a and d are internalized, stabilizing the structure, while positions b, c, e, f, and g are exposed on the surface of the protein

types of structures and focused on Coiled-coils (Fig. 11). Coiled-coil alpha helices can mediate oligomerization like beta-sheets. However, unlike beta-sheets, Coiled-coils can change the oligomeric state in response to signals from the environment. Coiled-coils have a periodic pattern of heptad repeats with hydrophobic residues in the a and d positions often occupied by leucine, isoleucine and valine. Moreover, glutamine can substitute for hydrophobic residues in the a and d positions. In Q-N rich prions including *Aplysia* CPEB, when two strands coil around each other, positions a and d internalize, stabilizing the structure, but positions b, c, e, f and g, which are usually occupied by polar/charged residues, are exposed on the surface of the protein and can be regulated, and they are, therefore, ideal candidate structures for regulated aggregate formation.

The Q-N rich prion domain of *Aplysia* CPEB, and that of its interactors, has a predictive pattern of hydrophobic residues similar to that found in Coiled-coil proteins (Fig. 12). Moreover, destabilization of Coiled-coils in these proteins by mutational replacements of glutamines impairs the aggregation of *Aplysia* CPEB

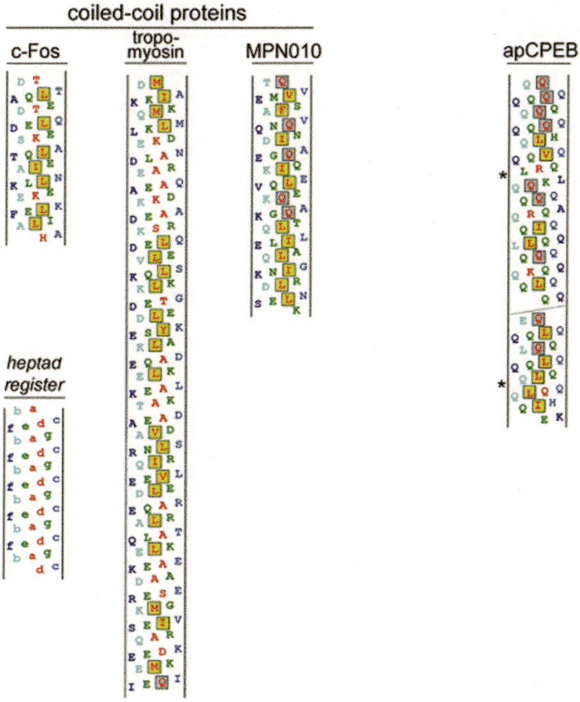

Fig. 12 The Q/N rich prion domain of ApCPEB has a predicted pattern of hydrophobic residues similar to that found in Coiled-coil (CC) regions of other proteins. Heptad repeats of a–d-spaced hydrophobic residues are highlighted in *yellow*. Q in a/d are highlighted in *gray*. Note the vertical alignment of a/d residues and the grouping of a/d hydrophobic residues in discrete clusters along the helices. *Asterisks* indicate stutters (Fiumara et al. 2010)

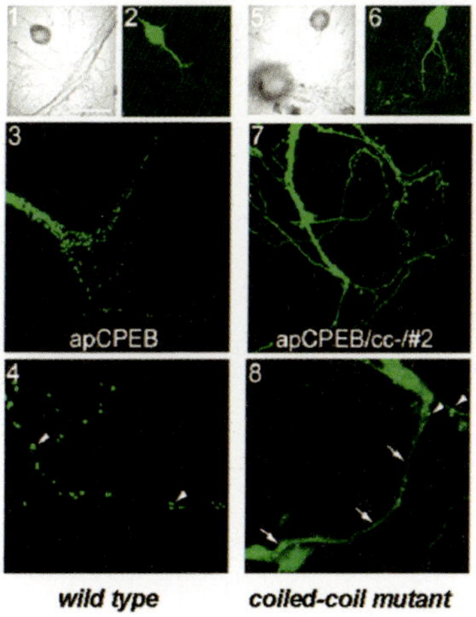

Fig. 13 Destabilization of Coiled-coil by mutational replacement of glutamines (apCPEB/cc/#2; mutant apCPEB) impairs aggregation of ApCPEB in vivo. *Arrowheads* indicate aggregates; *arrows* indicate cells with diffuse fluorescence (Fiumara et al. 2010)

in vivo (Fig. 13). Finally, Coiled-coils can exist on their own or as part of a regulated structural transition to beta-sheets (Fig. 14). There are two possibilities for the role of Coiled-coils in aggregation. One is that the whole aggregate is held together by Coiled-coil interactions. Alternatively, Coiled-coils can be important in

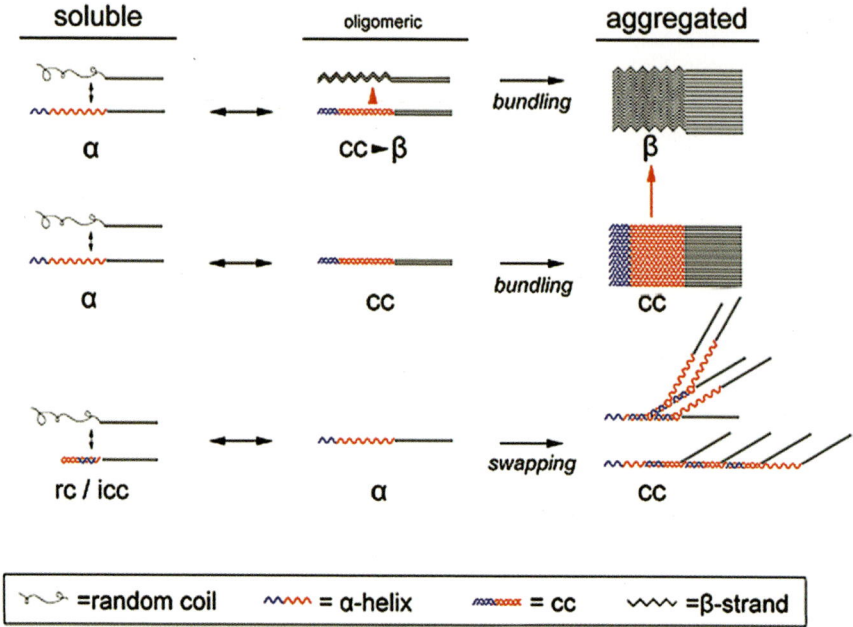

Fig. 14 Coiled-coils (CC) can exist on their own or as part of regulated structural transitions to beta sheets. Schematic representation of the possible role of CCs in the conformational dynamics of Q/N rich and polyQ proteins. Alpha-helices/CCs may represent intermediate structures facilitating the misfolding of helical protomers (*red arrowhead*) or CC fibers (*red arrow*) into β sheet polymers (*upper panel*) or self-sufficient mediators of conformational change and aggregation (*middle* and *lower panels*: icc, intramolecular CC; *gray bars*: other protein domains) (Fiumara et al. 2010)

initiating aggregation process, but at some point a beta strand-rich structure is established.

These Coiled-coils can serve as the scaffold to coordinate the translation of a population of interrelated messenger RNAs required for the stabilization of synaptic growth. In this way, CPEB aggregates can coordinate the polyadenylation of multiple messenger RNAs required for stabilization (Fig. 15).

An Overall View

The realization that protein conformational switches could provide a means for inheritance of phenotypes dates back more than 15 years (Wickner 1994). However, only a few proteins with this capacity have been reported in any system (Crow and Li 2011; Du et al. 2008; Shorter and Lindquist 2005). Most of these proteins have been found in the yeast *S. cerevisiae*, with the [*PSI*$^+$] element being the best understood. [*PSI*$^+$] is caused by an amyloid-like aggregated state of the

Fig. 15 Coiled-coils could serve as the scaffold to coordinate the transition of a population of interrelated mRNAs required for stabilization of synaptic growth (Raveendra et al. unpublished data)

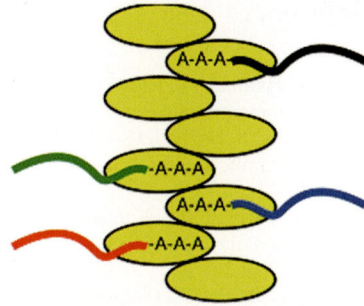

CPEB aggregate coordinate polyadenylation of multiple mRNAs required for stabilization

translation-termination factor Sup35p. In the prion conformation, the majority of Sup35p molecules are inactive, resulting in increased levels of nonsense suppression (Paushkin et al. 1996; Patino et al. 1996; Derkatch et al. 1996; Cox 1965) and programmed frame shifting (Namy et al. 2008), giving rise to RNA stability changes and functionally altered polypeptides and leading to phenotypes that can be advantageous under certain conditions (Eaglestone et al. 1999; True et al. 2004; True and Lindquist 2000; Halfmann et al. 2012). In an attempt to probe for prion domains, Lindquist and colleagues scanned the yeast genome bioinformatically for proteins with prion-like characteristics. They identified 24 new yeast proteins containing a prion-forming domain (PrD), thus expanding the repertoire of proteins that utilize prion-like conformational changes for their activity (Alberti et al. 2009).

Although prions were initially considered to be infectious proteinaceous agents that were associated with a class of fatal degenerative diseases of the mammalian brain, the discovery of fungal prions—which are not associated with disease— suggested the possibility that the effects of prion mechanisms on cellular physiology could be viewed in a different light. Fungal prions and epigenetic determinants could alter a *range* of cellular processes, including metabolism and gene expression pathways. These changes could lead to a variety of prion-associated phenotypes. The mechanistic similarities between prion propagation in snails, mammals and fungi suggest that prions are *not* a biological anomaly but, instead, could be a *newly* appreciated, and perhaps ubiquitous, regulatory mechanism. Our data have provided the first evidence for the existence of a prionic mechanism in the brain. We have found the presence of this mechanism in both the invertebrate and the mammalian brain, where it regulates and sustains local protein synthesis—a critical step in the stable, long-term maintenance of memory storage.

Acknowledgments Eric R. Kandel and Elias Pavlopoulos are supported by the Howard Hughes Medical Institute. Irina Derkatch is supported by National Institutes of Health grant 7 R01 GM070934.

References

Alberti S, Halfmann R, King O, Kapila A, Lindquist S (2009) A systematic survey identifies prions and illuminates sequence features of prionogenic proteins. Cell 137:146–158

Bally-Cuif L, Schatz WJ, Ho RK (1998) Characterization of the zebrafish Orb/CPEB-related RNA binding protein and localization of maternal components in the zebrafish oocyte. Mech Dev 77:31–47

Barco A, Alarcon JM, Kandel ER (2002) Expression of constitutively active CREB protein facilitates the late phase of long-term potentiation by enhancing synaptic capture. Cell 108:689–703

Casadio A, Martin KC, Giustetto M, Zhu H, Chen M, Bartsch D, Bailey CH, Kandel ER (1999) A transient, neuron-wide form of CREB-mediated long-term facilitation can be stabilized at specific synapses by local protein synthesis. Cell 99:221–237

Chang JS, Tan L, Wolf MR, Schedl P (2001) Functioning of the Drosophila orb gene in gurken mRNA localization and translation. Development 128:3169–3177

Coustou V, Deleu C, Saupe S, Begueret J (1997) The protein product of the het-s heterokaryon incompatibility gene of the fungus Podospora anserina behaves as a prion analog. Proc Natl Acad Sci USA 94:9773–9778

Cox BS (1965) ψ, a cytoplasmic suppressor of super-suppression in yeast. Heredity 20:505–521

Crow ET, Li L (2011) Newly identified prions in budding yeast, and their possible functions. Semin Cell Dev Biol 5:452–459

Crick F (1984) Memory and molecular turnover. Nature 312:101

Du Z, Park KW, Yu H, Fan Q, Li L (2008) Newly identified prion linked to the chromatin remodeling factor Swi1 in Saccharomyces cerevisiae. Nat Genet 40:460–465

Derkatch IL, Chernoff YO, Kushnirov VV, Inge-Vechtomov SG, Liebman SW (1996) Genesis and variability of ψ prion factors in Saccharomyces cerevisiae. Genetics 144(4):1375–1386

Dudek SM, Fields RD (2002) Somatic action potentials are sufficient for late-phase LTP-related cell signaling. Proc Natl Acad Sci USA 99:3962–3967

Eaglestone SS, Cox BS, Tuite MF (1999) Translation termination efficiency can be regulated in Saccharomyces cerevisiae by environmental stress through a prion-mediated mechanism. EMBO J 18:1974–1981

Fiumara F, Fioriti L, Kandel ER, Hendrickson WA (2010) Essential role of coiled coils for aggregation and activity of Q/N-rich prions and PolyQ proteins. Cell 143(7):1121–1135

Frey U, Morris RG (1997) Synaptic tagging and long-term potentiation. Nature 385:533–536

Frey U, Morris RG (1998) Weak before strong: dissociating synaptic tagging and plasticity-factor accounts of late-LTP. Neuropharmacology 37:545–552

Ganusova EE, Ozolins LN, Bhagat S, Newnam GP, Wegrzyn RD, Sherman MY, Chernoff YO (2006) Modulation of prion formation, aggregation, and toxicity by the actin cytoskeleton in yeast. Mol Cell Biol 26:617–629

Griffith JS (1967) Self-replication and scrapie. Nature 215:1043–1044

Groisman I, Jung MY, Sarkissian M, Cao Q, Richter JD (2002) Translational control of the embryonic cell cycle. Cell 109:473–483

Halfmann R, Jarosz DF, Jones SK, Chang A, Lancaster AK, Lindquist S (2012) Prions are a common mechanism for phenotypic inheritance in wild yeasts. Nature 482(7385):363–368

Hou F, Sun L, Zheng H, Skaug B, Jiang QX, Chen ZJ (2011) MAVS forms functional prion-like aggregates to activate and propagate antiviral innate immune response. Cell 146:448–461

Kandel ER (2001) The molecular biology of memory storage: a dialogue between genes and synapses. Science 294:1113–1120

Keleman K, Krüttner S, Alenius M, Dickson BJ (2007) Function of the Drosophila CPEB protein Orb2 in long-term courtship memory. Nat Neurosci 10:1587–1593

Lisman J, Malenka RC, Nicoll RA, Malinow R (1997) Learning mechanisms: the case for CaM-KII. Science 276:2001–2002

Majumdar A, Cesario WC, White-Grindley E, Jiang H, Ren F, Khan MR, Li L, Choi EM, Kannan K, Guo F, Unruh J, Slaughter B, Si K (2012) Critical role of amyloid-like oligomers of Drosophila Orb2 in the persistence of memory. Cell 148:515–529

Martin KC, Casadio A, Zhu H, Yaping E, Rose JC, Chen M, Bailey CH, Kandel ER (1997) Synapse-specific, long-term facilitation of aplysia sensory to motor synapses: a function for local protein synthesis in memory storage. Cell 91:927–938

Mastushita-Sakai T, White-Grindley E, Samuelson J, Seidel C, Si K (2010) Drosophila Orb2 targets genes involved in neuronal growth, synapse formation, and protein turnover. Proc Natl Acad Sci USA 107:11987–11992

Mendez R, Murthy KG, Ryan K, Manley JL, Richter JD (2000) Phosphorylation of CPEB by Eg2 mediates the recruitment of CPSF into an active cytoplasmic polyadenylation complex. Mol Cell 2000:1253–1259

Miniaci MC, Kim JH, Puthanveettil SV, Si K, Zhu H, Kandel ER, Bailey CH (2008) Sustained CPEB-dependent local protein synthesis is required to stabilize synaptic growth for persistence of long-term facilitation in Aplysia. Neuron 59:1024–1036

Namy O, Galopier A, Martini C, Matsufuj S, Fabret C, Rousset JP (2008) Epigenetic control of polyamines by the prion [PSI(+)]. Nat Cell Biol 10:1069–1075

Patino MM, Liu JJ, Glover JR, Lindquist S (1996) Support for the prion hypothesis for inheritance of a phenotypic trait in yeast. Science 273:622–626

Paushkin SV, Kushnirov VV, Smirnov VN, Ter-Avanesyan MD (1996) Propagation of the yeast prion-like [psi+] determinant is mediated by oligomerization of the SUP35-encoded polypeptide chain release factor. EMBO J 15(12):3127–3134

Pavlopoulos E, Trifilieff P, Chevaleyre V, Fioriti L, Zairis S, Pagano A, Malleret G, Kandel ER (2011) Neuralized1 activates CPEB3: a function of non-proteolytic ubiquitin in synaptic plasticity and memory storage. Cell 147:1369–1383

Prusiner SB (1982) Novel proteinaceous infectious particles cause scrapie. Science 216:136–144

Puthanveettil S, Kandel ER (2011) Molecular mechanisms for the initiation and maintenance of long-term memory storage. In: Curran T, Christen Y (eds) Two faces of evil: cancer and neurodegeneration. Springer, New York, pp 143–160

Richter JD (1999) Cytoplasmic polyadenylation in development and beyond. Microbiol Mol Biol Rev 63:446–456

Schroeder KE, Condic ML, Eisenberg LM, Yost HJ (1999) Spatially regulated translation in embryos: asymmetric expression of maternal Wnt-11 along the dorsal-ventral axis in Xenopus. Dev Biol 214:288–297

Shorter J, Lindquist S (2005) Prions as adaptive conduits of memory and inheritance. Nat Rev Genet 6:435–450

Si K, Giustetto M, Etkin A, Hsu R, Janisiewicz AM, Miniaci MC, Kim JH, Zhu H, Kandel ER (2003a) A neuronal isoform of CPEB regulates local protein synthesis and stabilizes synapse specific long-term facilitation in aplysia. Cell 115:893–904

Si K, Lindquist S, Kandel ER (2003b) A neuronal isoform of the aplysia CPEB has prion-like properties. Cell 115:879–891

Si K, Choi YB, White-Grindley E, Majumdar A, Kandel ER (2010) Aplysia CPEB can form prion-like multimers in sensory neurons that contribute to long-term facilitation. Cell 140:421–435

Stebbins-Boaz B, Hake LE, Richter JD (1996) CPEB controls the cytoplasmic polyadenylation of cyclin, Cdk2 and c-mos mRNAs and is necessary for oocyte maturation in Xenopus. EMBO J 15:2582–2592

Tan L, Chang JS, Costa A, Schedl P (2001) An autoregulatory feedback loop directs the localized expression of the Drosophila CPEB protein Orb in the developing oocyte. Development 128:1159–1169

Theis M, Si K, Kandel ER (2003) Two previously undescribed members of the mouse CPEB family of genes and their inducible expression in the principal cell layers of the hippocampus. Proc Natl Acad Sci USA 100(16):9602–9607

Thompson KR, Otis KO, Chen DY, Zhao Y, O'Dell TJ, Martin KC (2004) Synapse to nucleus signaling during long-term synaptic plasticity; a role for the classical active nuclear import pathway. Neuron 44:997–1009

True HL, Lindquist SL (2000) A yeast prion provides a mechanism for genetic variation and phenotypic diversity. Nature 407:477–483

True HL, Berlin I, Lindquist SL (2004) Epigenetic regulation of translation reveals hidden genetic variation to produce complex traits. Nature 431:184–187

Wickner RB (1994) [URE3] as an altered URE2 protein: evidence for a prion analog in Saccharomyces cerevisiae. Science 264:566–569

Index

M. Jucker and Y. Christen (eds.), *Proteopathic Seeds and Neurodegenerative Diseases*, 153
Research and Perspectives in Alzheimer's Disease,
DOI 10.1007/978-3-642-35491-5, © Springer-Verlag Berlin Heidelberg 2013